Contents

Unit 17	Compare Numbers	1
Unit 18	Ten and Ones	19
Unit 19	Numbers to 20	25
Unit 20	Number Bonds	31
Unit 21	Addition	41
Unit 22	Counting On	53
Unit 23	Subtraction	65
Unit 24	Part	73
Unit 25	Counting Back	79
Unit 26	Addition and Subtraction	85
Unit 27	Numbers to 40	101
Unit 28	Ordering	119
Unit 29	Time	135
Unit 30	Numbers to 100	141
Unit 31	Even/Odd	147
Unit 32	Fractions	153

Unit 17 — Compare Numbers

Draw a set with one more.

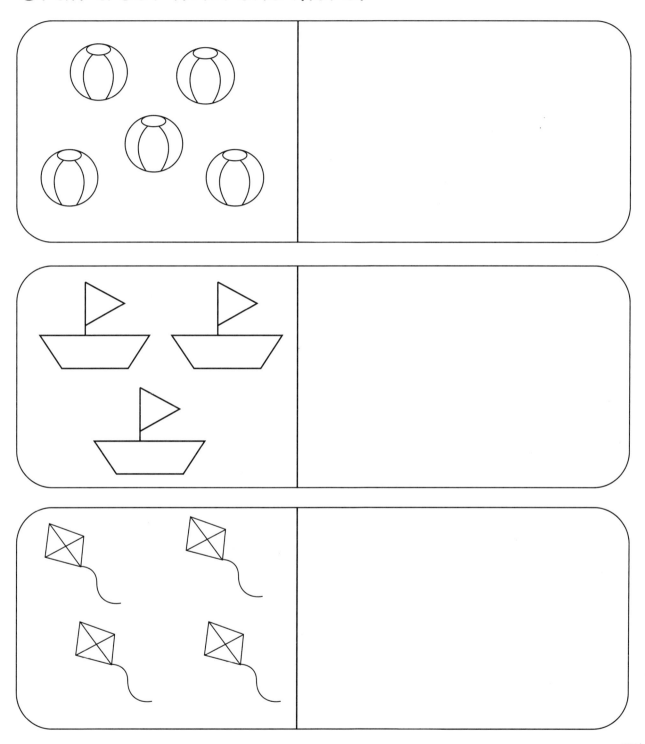

Concept: Find the number that is 1 more.
Introduction: Use linking cubes. Give the child numeral cards 1 through 10 and have him/her put them in order. Point to one number, e.g., 4, and have him/her count out some cubes and link them together. Then ask the child to put one more cube on the tower. Ask, "What is 1 more than 4?" Draw some objects on the board. Ask the child to count them. Then ask the child to tell you what number is 1 more. Add another object to the drawing.

Unit 17 — Compare Numbers

Circle the set that has one item more than the picture on the left.

2

Concept: Compare sets to find the one that is 1 more.
Introduction: Give the child a set of up to 9 objects. Ask the child to count the objects and write the number. Then ask, "How many will there be if I add one more?" Have the child write the new number.
Using This Page: Be sure the child understands that he/she is to circle the set that has one more than the one on the left. Guide him/her in doing the first one, or show an example on the board.

Unit 17 — Compare Numbers

Fill in the blanks.

1 more than 3 is _____ .

1 more than 6 is _____ .

1 more than 9 is _____ .

Concept: Find the number that is 1 more.
Introduction: Give the child some counters. Write a number and ask the child to count out the corresponding number of counters. Then ask the child to add one more and tell you how many counters there are now. Write, "1 more than ___ is ___." Repeat with other numbers within 10.

Unit 17 — Compare Numbers

Circle the number that is one more than the number on the left. The first one has been done for you.

8	⑨	5	4
6	3	4	7
4	5	4	7
9	10	8	9
5	8	6	9
7	5	8	6

Concept: Find the number that is 1 more.
Introduction: Call out a number within 10. Ask the child to tell you the number that is 1 more. Repeat.

Unit 17 — Compare Numbers

Draw a set with one less.

Concept: Find the number that is one less.
Introduction: Give the child the numeral cards 1 through 10 and ask him/her to put them in order. Point to the last one and have him/her say the number. Then point to the next-to-last one and have him/her say the number. Continue down to 1. Draw some objects on the board. Ask the child to count them. Then erase one object. Ask the child to tell you what number is 1 less.

Unit 17 — Compare Numbers

Write how many. Circle the set that has 1 less than the picture on the left.

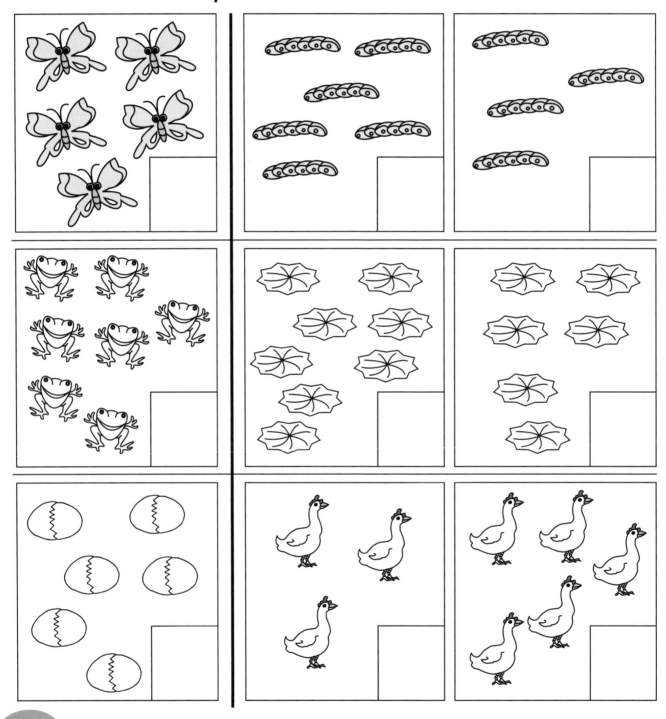

Unit 17 — Compare Numbers

Cross one out and fill in the blank. The first one is done for you.

1 less than 4 is __3__ .

1 less than 6 is _____ .

1 less than 5 is _____ .

1 less than 9 is _____ .

Concept: Find the number that is 1 less.
Introduction: Give the child some objects such as counters. Write a number and ask the child to count out the corresponding number of objects. Then ask the child to remove one and tell you how many counters there are now. Write, "1 less than ___ is ___." Fill in the first blank and ask the child to fill in the second blank. Repeat with other numbers within 10.

Unit 17 — Compare Numbers

Circle the number that is one **less** than the number on the left. The first one has been done for you.

8	9	5	⦿7
6	5	4	3
4	8	3	2
9	10	8	9
5	4	6	9
7	5	8	6

Concept: Find the number that is 1 less.
Introduction: Call out a number within 10. Ask the child to tell you the number that is 1 less. Repeat with other numbers within 10.

Unit 17 — Compare Numbers

Count and write the numbers. Circle the number that is **greater**.

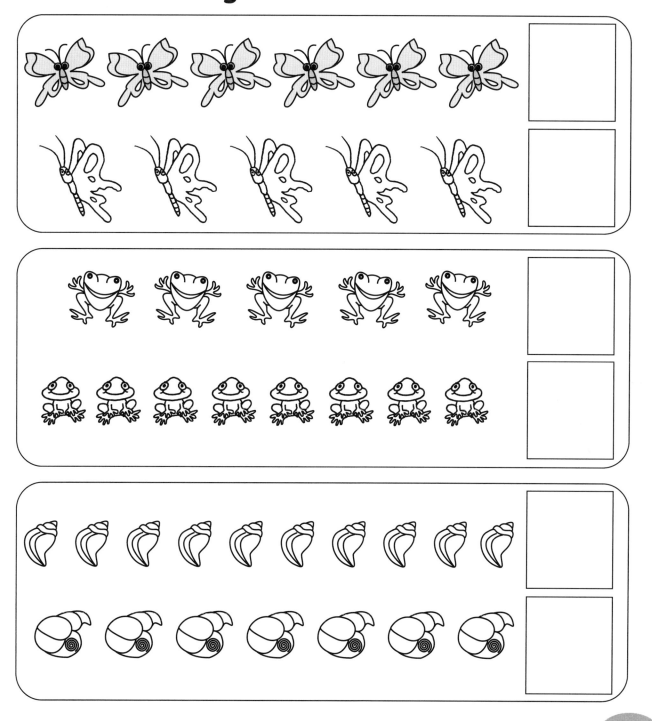

Concept: Compare quantities.
Introduction: Display two sets of objects of unequal amounts. Ask the child to count the number in each set and write down the numbers. Then ask the child to pair an item from one set with an item from another set to determine which set has more. Point to the numbers and ask the student which number is more, or **greater**, than the other number.

Unit 17 — Compare Numbers

Color the number in each shape that is greater.

Concept: Compare Numbers.
Introduction: Use linking cubes. Have the child link them to create towers for 1 through 10 and have him/her put them in order. Give the child numeral cards for 1 through 10 and ask him/her to put them in order next to the towers. Point to two numbers and ask which one is greater.

Unit 17 — Compare Numbers

Count and write the numbers. Circle the number that is smaller.

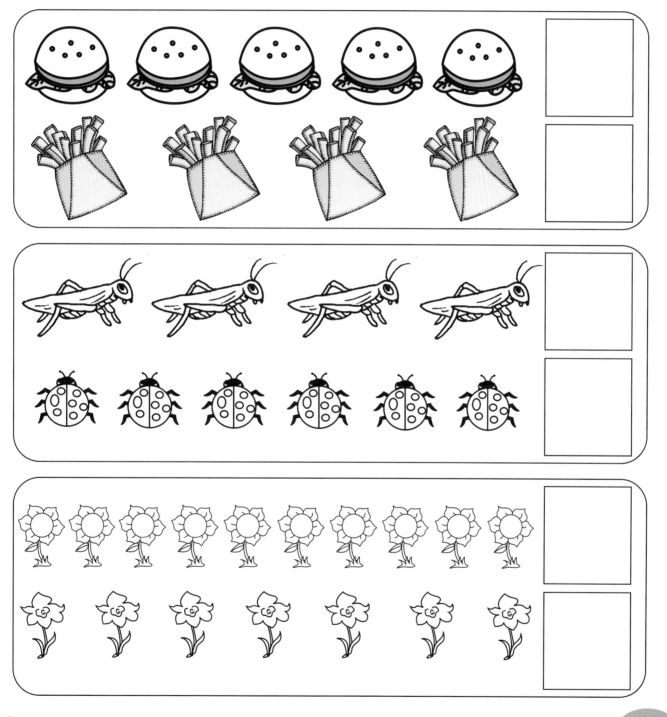

Concept: Compare quantities.
Introduction: Display two sets of objects of unequal amounts. Ask the child to count the number in each set and write down the numbers. Then ask the child to pair an item from one set with an item from another set to determine which set has less. Point to the numbers and ask the student which number is less, or **smaller**, than the other number.

Unit 17 — Compare Numbers

Color the number in each shape that is smaller.

Concept: Compare numbers.
Introduction: Give the child number cards for 1 through 10 and ask the child to put them in order. Point to two numbers and ask which one is smaller.

Unit 17 — Compare Numbers

Are there more boys or more kites? Circle the correct answer. Then fill in the blank with how many more.

There are more kites than boys.

There are more boys than kites.

There are _____ **more**.

Concept: Compare numbers.
Introduction: Give the child 2 sets of objects. Ask him/her to point out the set with more items. Then, ask him/her to find out how many more the larger set has. The child can pair the items from one set with those form the other and count the leftovers. Get the child to say, "____ is ____ more than ____."

Unit 17   Compare Numbers

Are there less balloons or less children? Circle the correct answer. Then fill in the blank with how many less.

There are less balloons than children.

There are less children than balloons.

There are _____ less.

Concept: Compare numbers.
Introduction: Give the child 2 sets of objects. Ask him/her to point out the set with less items. Then, ask the child to find out how many less the smaller set has. The child can pair the items from one set with those form the other and count the leftovers. Get the child to say, "____ is ____ less than ____."

Unit 17 — Compare Numbers

Look at the pictures and write the correct numbers in each box.

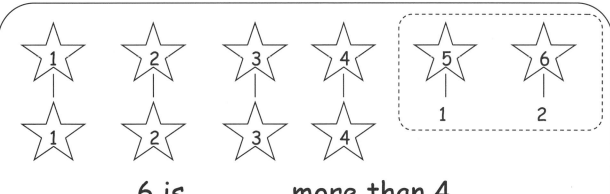

6 is _____ more than 4.

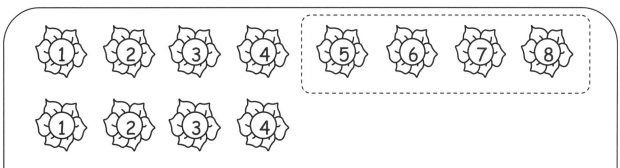

8 is _____ more than 4.

9 is 3 more than _____.

Concept: Compare numbers.
Introduction: Write down two numbers within 10. Have the child use objects such as counters to find out which number is more and how much more.

Unit 17 — Compare Numbers

Look at the pictures and write the correct numbers in each box.

5 is _____ less than 7.

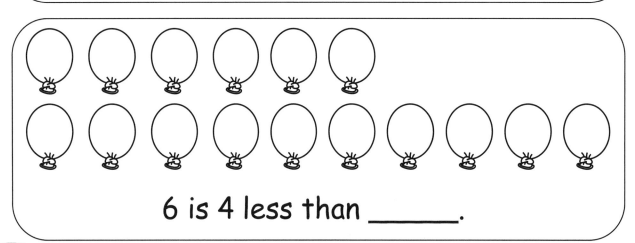

3 is _____ less than 6.

6 is 4 less than _____.

Concept: Compare numbers.
Introduction: Write down two numbers within 10. Have the child use objects such as counters to find out which number is less and how much less.

Unit 17 — Compare Numbers

Draw 2 more.

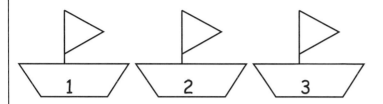

_____ is 2 more than 3.

Draw 3 more.

_____ is 3 more than 2.

Cross off 5.

_____ is 5 less than 8.

Concept: Compare numbers.
Introduction: Display a set of 5 objects. Ask the child what is 3 more than 5. Have the child add 3 more objects and tell you the answer. Display a set of 7 objects. Ask the child what is 2 less than 7. Have the child take away 2 objects and tell you the answer.
Using this page: Help the child with the instructions for each problem on this page.

Unit 17 — Compare Numbers

Write the numbers. Read the sentences.

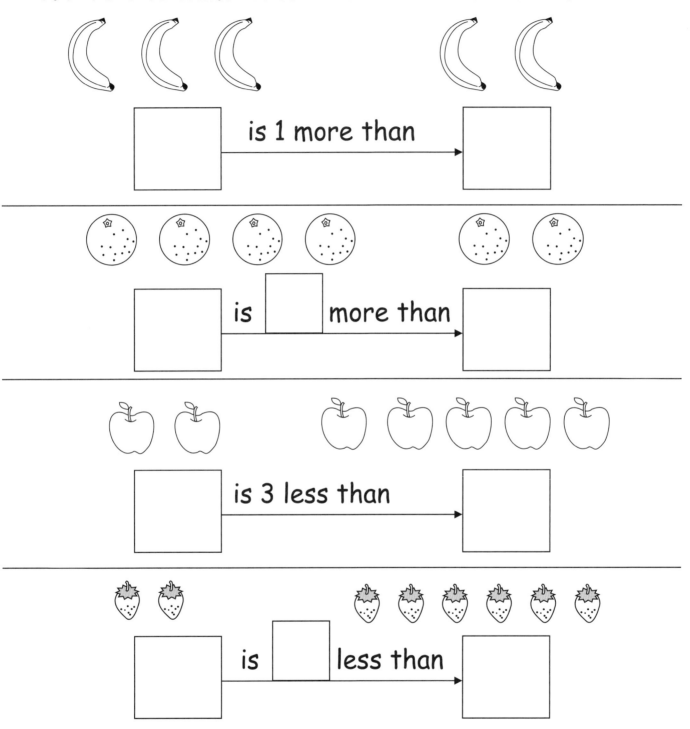

Concept: Compare numbers.
Introduction: Display the numeral cards 1-6 in order. Have the numeral cards for 7-10 available. Ask the child for the number that is 2 more than 6. Have him/her add the numerals 7 and 8 to find the answer. Then ask the child for the number that is 3 less than 8. Have him/her remove the cards 8, 7, and 6 to find the answer.

Unit 18 — Tens and Ones

Color the sets that have 10 objects.

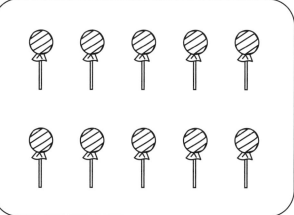

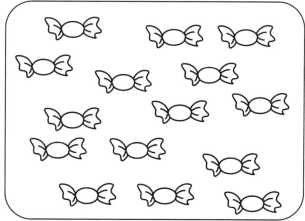

Concept: Count to ten when there are more than ten objects.
Introduction: Show the child a set of 10-20 objects. Ask the child to make a group of ten and a group of leftovers.

Unit 18 — Tens and Ones

Color 10 objects. Write the numbers.

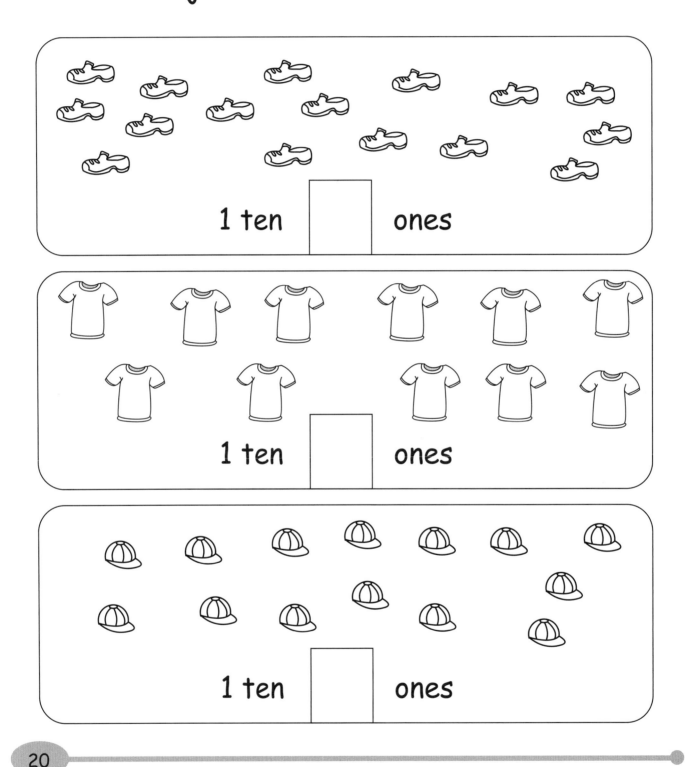

Unit 18 — Tens and Ones

Count the tens and ones. Write the ones and the number.

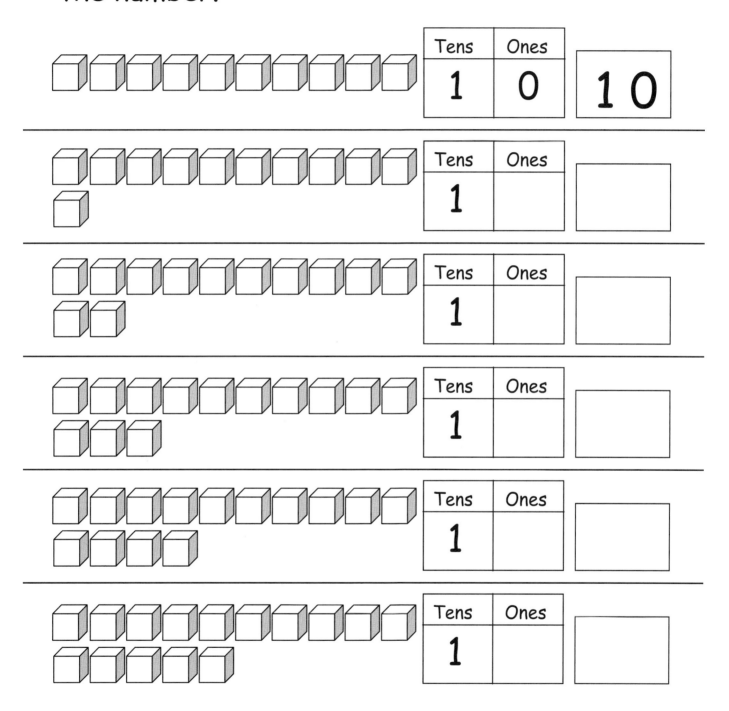

Concept: Count to 15 and recognize the numerals.
Introduction: Use linking cubes to make a linked set of ten. Display the ten and ask the child to count. Write "10" and tell the child there is 1 ten and no ones, pointing to the digits. Add one more object and say "eleven". Continue with 12 through 15. Get the child to count other sets of objects to 15.
One Step Further: Show the child a numeral card between 10 and 15 and ask the child to read it. Repeat with other numbers between 10 and 15.

Unit 18 — Tens and Ones

Count the tens and ones. Write the tens, ones, and the number.

Tens	Ones

Tens	Ones

Tens	Ones

Tens	Ones

Tens	Ones

Tens	Ones

Concept: Count to 20 and recognize the numerals.
Introduction: Use linking cubes to make a linked set of ten. Display the ten and five ones and ask the child to count. Add 1 more, saying, "sixteen". Continue with 17 through 20. Have the child count the ones when you add one more to 19 to make 20. Point out that they make another ten, and link them together. Get the child to count other sets of objects to 20.
One Step Further: Show the child a numeral card between 10 and 20 and ask the child to read it. Repeat with other numbers between 10 and 20.

Unit 18 — Tens and Ones

Count. Write the number.

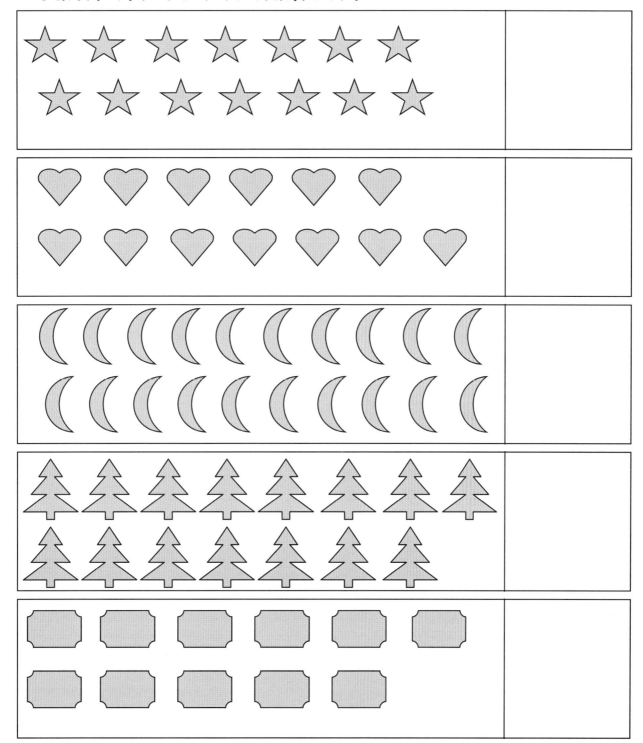

Unit 18 Tens and Ones

Draw more to make the number.

18

15

12

16

Concept: Count to 20 and write the numerals.
Introduction: Give the child a numeral card between 10 and 20 and ask him/her to count out the correct number of objects.

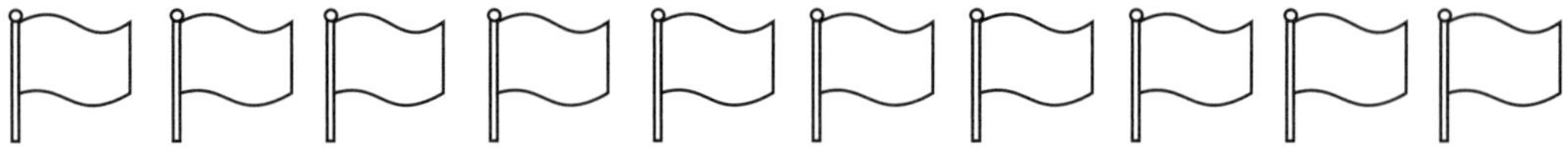

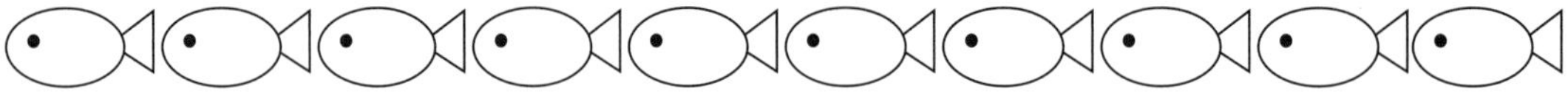

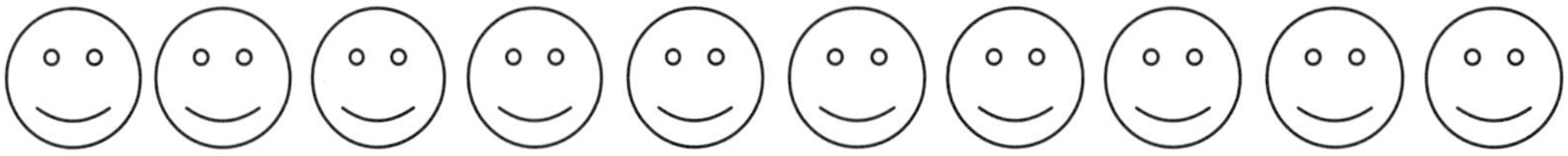

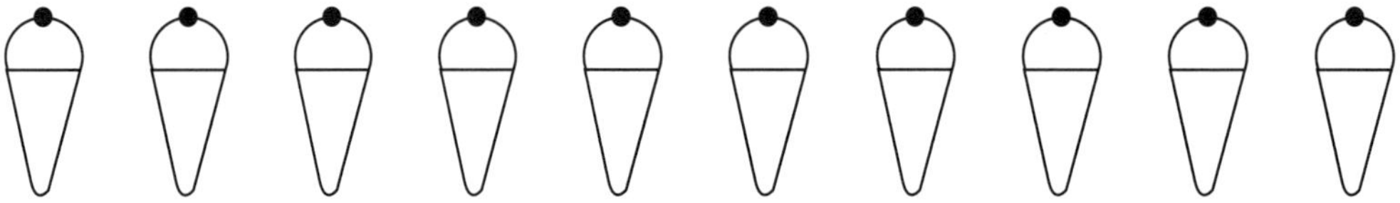

Unit 19 — Numbers to 20

Color more rectangles and fill in the missing numbers.

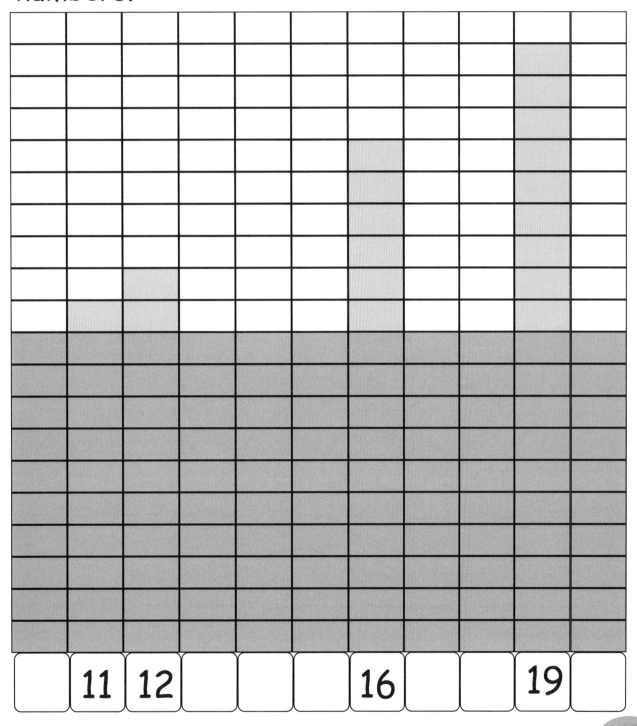

| | 11 | 12 | | | 16 | | | 19 | |

Concept: Numbers have order.
Introduction: Give the child numeral cards 10-20 and ask him/her put them in order. Ask questions such as, "What number is 1 more than 16? What comes before 18?"

Unit 19 — Numbers to 20

Join the dots in order according to the numbers. What picture do you see?

Concept: Numbers have order.

Unit 19 — Numbers to 20

Join the dots in order according to the numbers. What picture do you see?

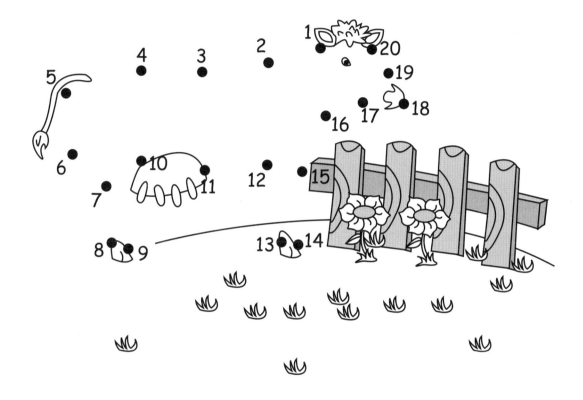

Concept: Numbers have order.
Introduction: Give the child numeral cards 1-20 and ask him/her to put them in order. Have the child name each one.

Unit 19 — Numbers to 20

Write the missing numbers.

Unit 19 — Numbers to 20

Fill in the missing numbers, starting at 11.

Concept: Write numbers to 20.
Introduction: Put the numeral cards 10-20 in order. Remove some, mix them up, and have the child replace them.

Unit 19 — Numbers to 20

Fill in the missing numbers.

8 9 ▢ ▢ 12

15 16 ▢ ▢ ▢

▢ 13 14 ▢ ▢

14 ▢ ▢ 17 ▢

▢ 17 ▢ ▢ 20

Concept: Write numbers to 20.
Introduction: Tell the child one number between 11 and 20 and ask him/her to count on from that number. Repeat with another number.
One Step Further: Teach the child to count backwards from 20.

Unit 20 — Number Bonds

There are 2 boxes of toys. How many toys are there altogether? Fill in the correct numbers.

3 and 2 make ☐

Concept: Find the total for two sets.
Introduction: Display two sets of objects and get the child to count the number in each set and write the numbers. Then put the two sets together and ask the child to find the total number. Ask how many there are altogether. Get the child to say, "___ and ___ make ___."

Unit 20 — Number Bonds

There are 2 cages of animals. How many animals are there altogether? Fill in the correct numbers.

 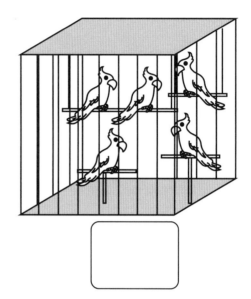

☐ and ☐

make 9

Unit 20 — Number Bonds

Write the matching numbers in the circles.

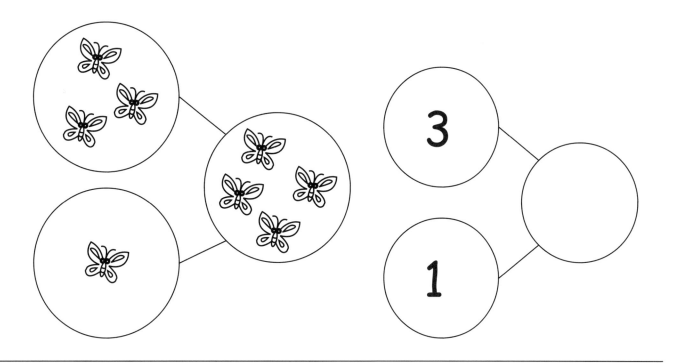

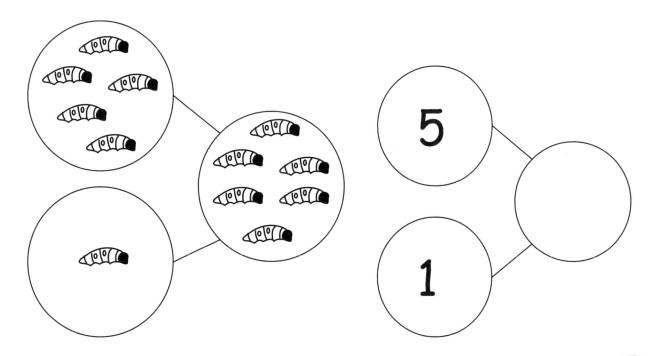

Concept: Build number bonds.
Introduction: Display two sets of objects and ask the child to find the amount in each set. Draw two circles and write the numerals. Then ask the child to find how many objects there are altogether. Write the total in a third circle and draw lines from the two parts to the total. Point to the parts, and then the total, saying, "___ and ___ make ___."

Unit 20 — Number Bonds

Write the matching numbers in the circles.

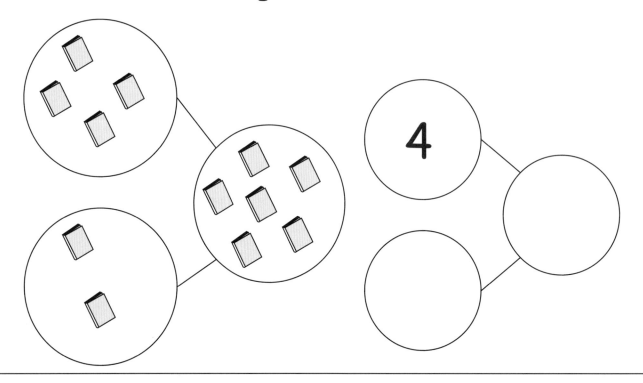

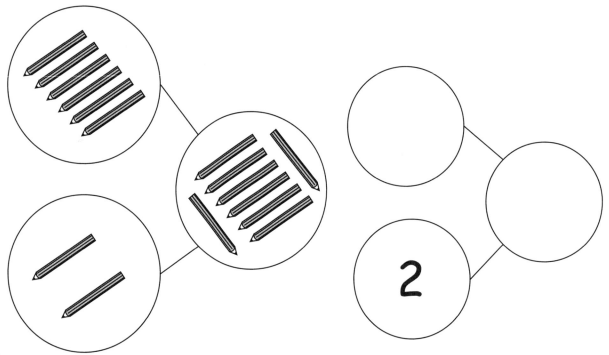

Unit 20 — Number Bonds

Write the matching numbers in the circles.

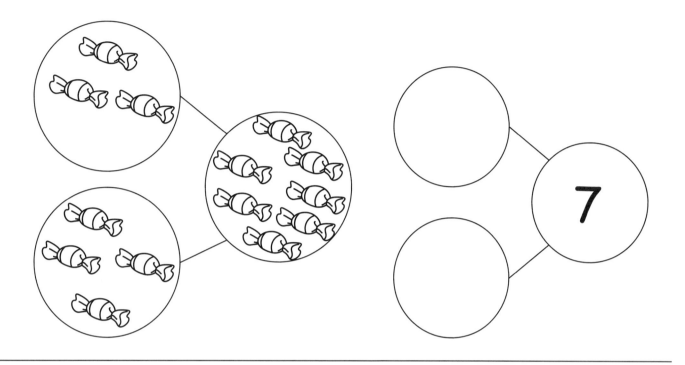

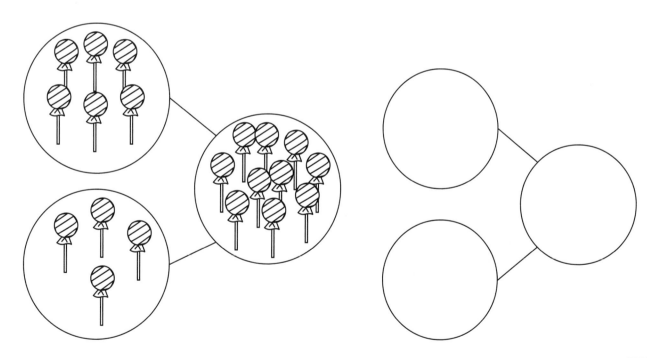

Unit 20 — Number Bonds

Write the matching numbers in the circles.

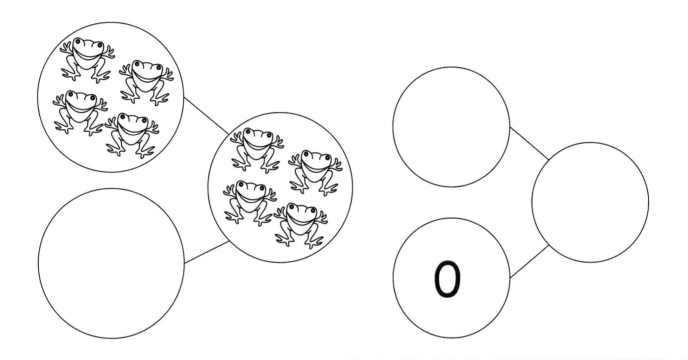

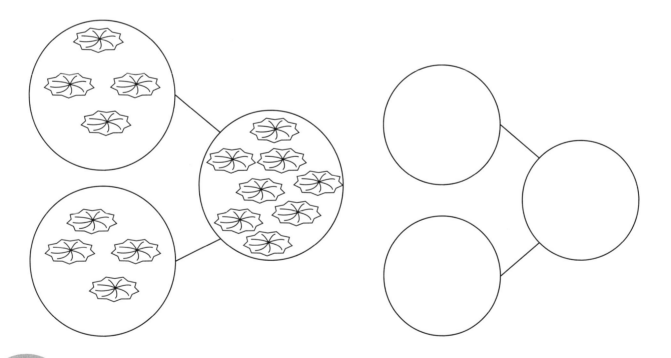

Unit 20 — Number Bonds

Color the objects with two different colors. Fill in the number bond.

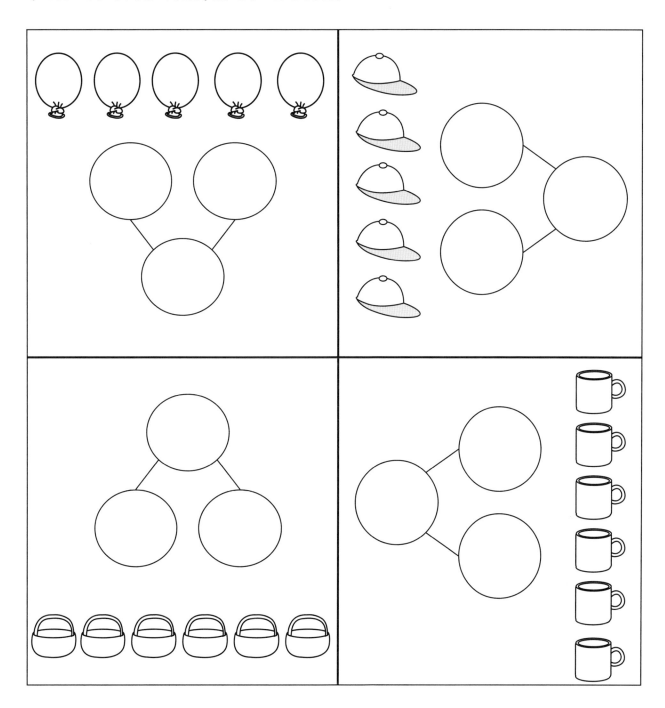

Concept: Build number bonds for 5 and 6.
Introduction: Give the child 5 linking cubes, linked. Ask the child to separate the cubes into two sets in various ways to find different ways to make 5. Help the child fill in number bonds for each. Guide the student in deciding which circle is for the total, and which are the parts. Repeat with 6 linking cubes.

Unit 20 — Number Bonds

Color the blocks with two different colors. Fill in the number bond.

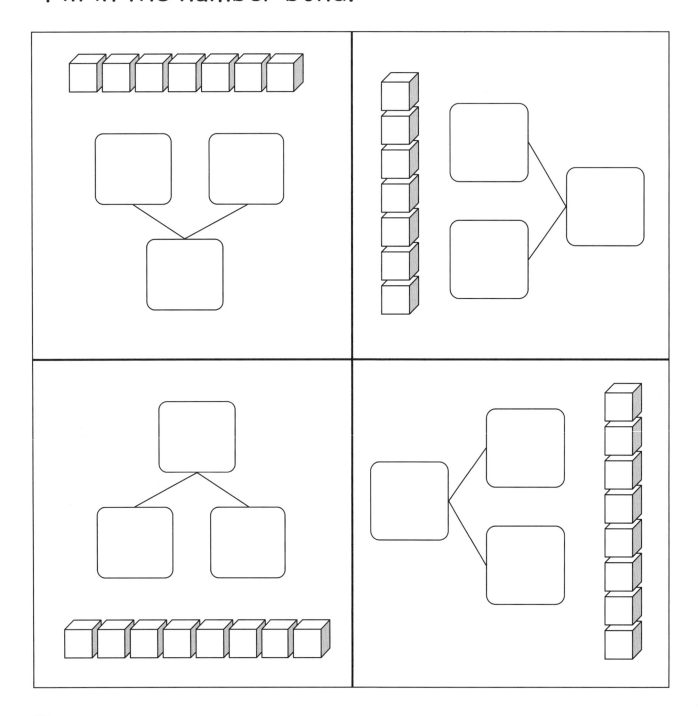

Concept: Build number bonds for 7 and 8.
Introduction: Give the child 7 linking cubes, linked. Ask the child to separate the cubes into two sets in various ways to find different ways to make 7. Help the child fill in number bonds for each way. Repeat with 8 linking cubes.

Unit 20 — Number Bonds

Draw lines to make two sets in different ways. The first one has been started for you. Fill in the number bonds.

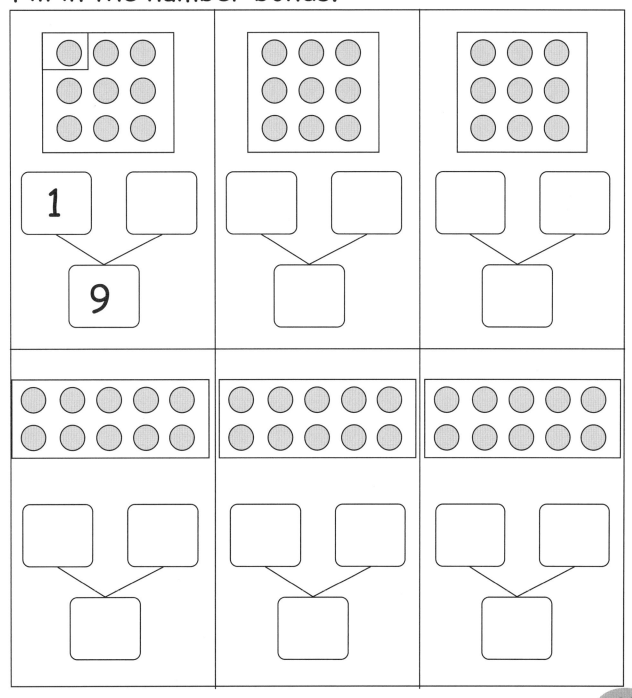

Unit 20 — Number Bonds

Fill in the number bonds.

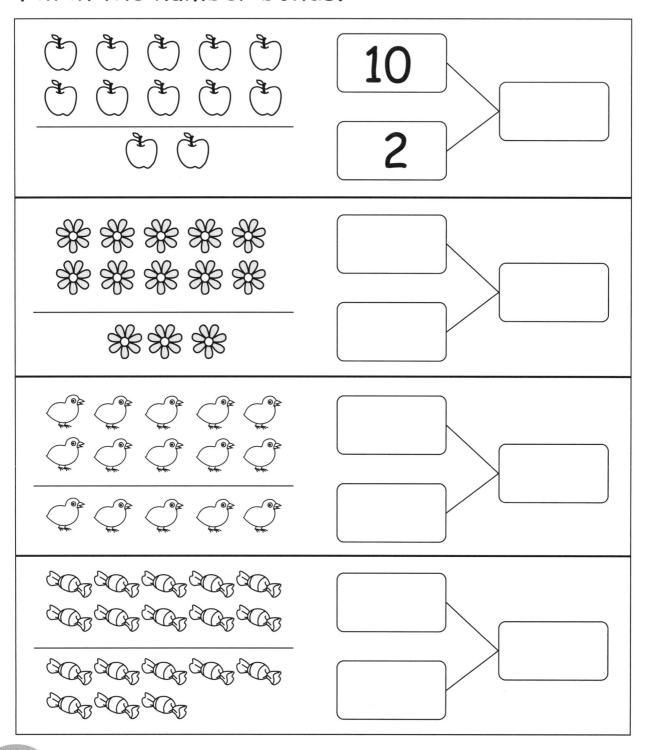

Concept: Complete number bonds for ten and ones.
Introduction: Give the child between 11 and 19 objects, such as 15, and have the child count and tell you the number by making a group of ten first. Say, "10 and 5 make 15" and draw a number bond. Repeat with other numbers between 10 and 19.

Unit 21 — Addition

Count the total number of objects in each set and write the number. The first one has been done for you.

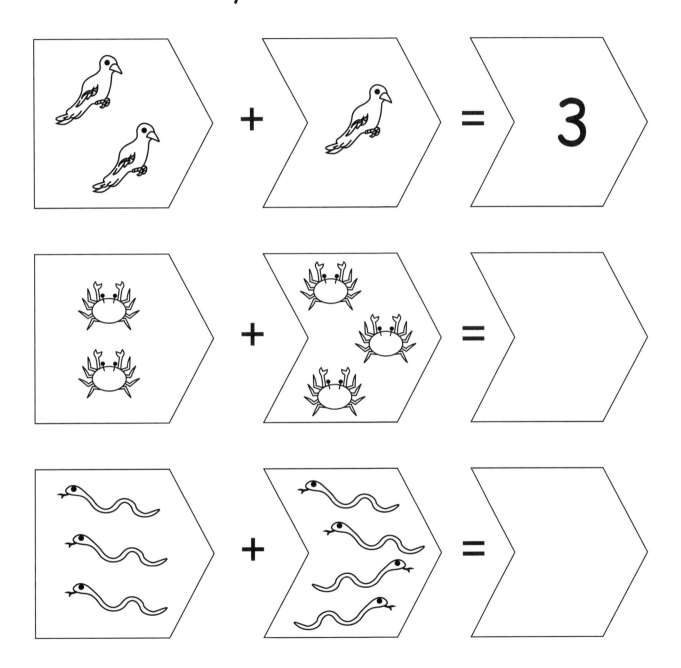

Concept: Understand simple addition problems.
Introduction: Display two sets of objects, such as 3 and 6, and ask the child how many there are altogether. Write "3 + 6 = 9". Read it as, "3 and 6 make 9." Point to the "+" symbol and say we use it to mean "and". Point to the "=" symbol and say we use it to mean "makes".

Unit 21 — Addition

Count the total number of objects in each set and write the number.

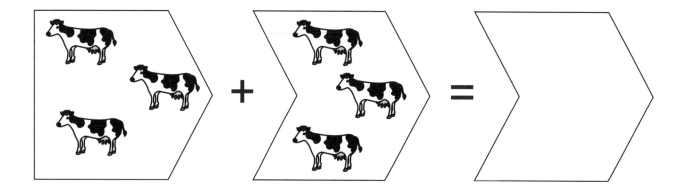

Concept: Understand simple addition problems.

Unit 21 — Addition

Count the number of sea creatures. Fill in the numbers.

Number of fish ☐   Number of starfish ☐

Total number of sea creatures: ☐

6 + 3 = ☐

Concept: Relate addition equations to number bonds.
Introduction: Write a number bond. Have the child use counters or other objects for each part and to find the total. Write the addition equation. Point to the parts of the number bond, saying "___ and ___ make ___" and then do the same for the addition equation.

Unit 21 — Addition

How many cars are there altogether? Write the numbers.

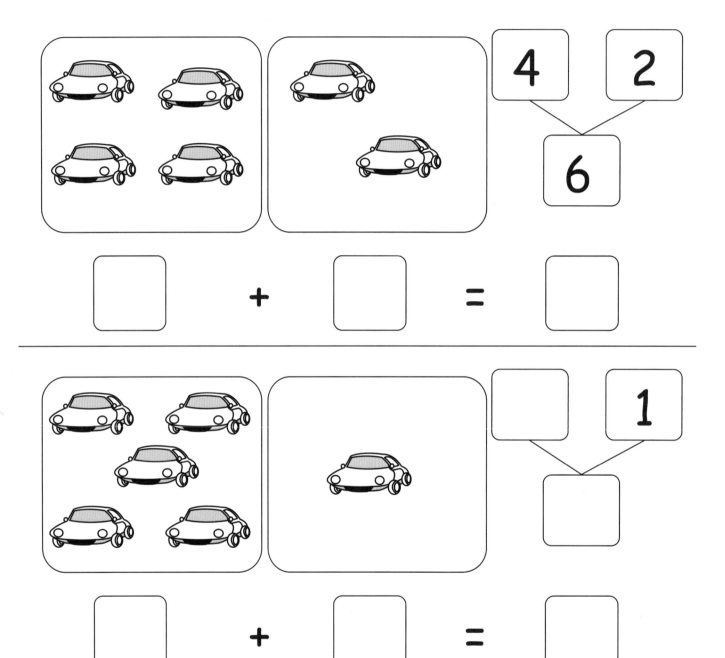

Concept: Relate addition equations to number bonds.
One Step Further: Tell the child that "+" can be read as "plus" and "=" can be read as "equals". Tell the child that "equals" means "is the same as". Write an addition equation, such as 2 + 1 = 3, and have the child read it as, "Two plus one equals three."

Unit 21 — Addition

Add. Write the numbers.

3 + 1 = _____

2 + 7 = _____

5 + 5 = _____

8 + 3 = _____

Concept: Understand simple addition problems.
Introduction: Tell the child that when we find our how many there are altogether, we 'add'. Display two sets of objects and ask the child to add them. Write the addition sentence and ask the child which two numbers are being added together.

Unit 21 — Addition

Add. Write the numbers.

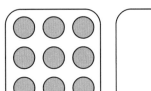 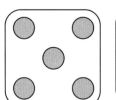

9 + 1 = _____   5 + 6 = _____

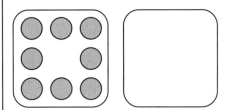 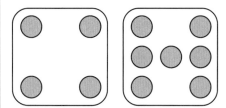

8 + 0 = _____   4 + 7 = _____

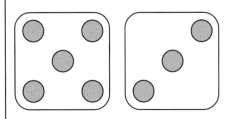 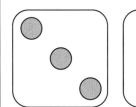

5 + 3 = _____   3 + 0 = _____

Concept: Understand simple addition problems.
Introduction: Write some addition expressions and have the student use counters to find the answers. Include some where 0 is added.

Unit 21 — Addition

Write how many blocks there are altogether.

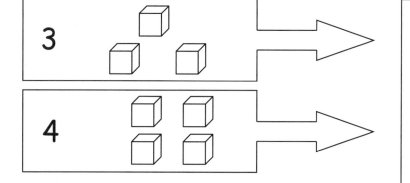

$$\begin{array}{r}3\\+\,4\\\hline\end{array}$$

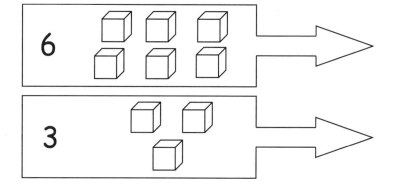
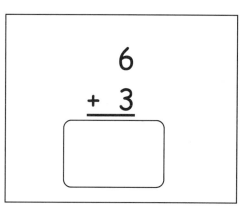

$$\begin{array}{r}6\\+\,3\\\hline\end{array}$$

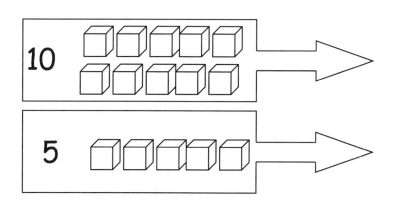

$$\begin{array}{r}10\\+\,5\\\hline\end{array}$$

Concept: Understand the vertical representation of an addition equation.
Introduction: Write an addition expression, such as 4 + 2, and have the child find the answer using counters. Write the answer, 4 + 2 = 6. Then, rewrite the problem vertically. Tell the child that this is another way that we can write down the numbers we are adding. Instead of an equal sign, we write a line under the two numbers we are adding to separate them from the total.

Unit 21 — Addition

Add. Write the numbers.

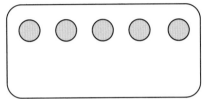
5
+ 2

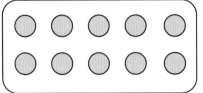
10
+ 4

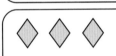
10
+ 3

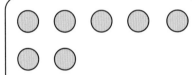
7
+ 3

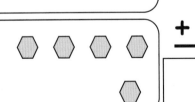
6
+ 5

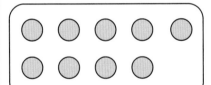
9
+ 4

Concept: Understand simple addition problems.

Unit 21 — Addition

Color the sets with the answer 7.

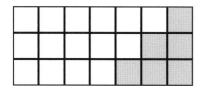 **7**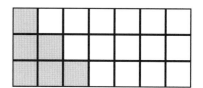

3 + 4	1 + 6
9 + 1	4 + 3
6 + 1	2 + 5
6 + 6	3 + 5
2 + 4	5 + 2

Concept: Understand different combinations of number bonds.
Introduction: Use two colors of linking cubes. Get the child to link the two colors in different ways to make 7 (1 and 6, 2 and 5, 3 and 4). Write the addition equations for the different ways to make 7. Turn each set of linked cubes around to show that either number can come first, e.g.; 6 + 1 is the same as 1 + 6.

Unit 21 — Addition

Color the sets with the answer 8.

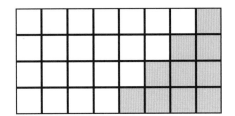　8　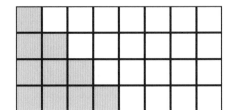

3 + 4	2 + 6
7 + 1	4 + 4
5 + 3	2 + 5
6 + 6	3 + 5
6 + 2	5 + 2

Concept: Understand different combinations of number bonds.
Introduction: Use 2 colors of linking cubes. Get the child to link the two colors in different ways to make 8 (1 and 7, 2 and 6, 3 and 5, 4 and 4). Write the addition equations for the different ways to make 8. Turn each set of linked cubes around to show that either number can come first, e.g., 6 + 2 is the same as 2 + 6.

Unit 21 — Addition

Fill in the numbers with all the ways to make 9.

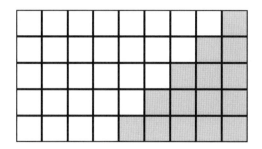 9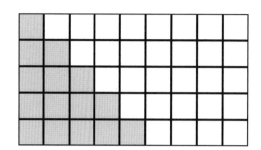

8 + 1 = 9   1 + ☐ = 9

7 + ☐ = 9   2 + ☐ = 9

6 + ☐ = 9   3 + ☐ = 9

5 + ☐ = 9   4 + ☐ = 9

Concept: Understand different combinations of number bonds.
Introduction: Use 2 colors of linking cubes. Get the child to link the two colors in different ways to make 9 (1 and 8, 2 and 7, 3 and 6, 5 and 4). Write the addition equations for the different ways to make 9. Turn each set of linked cubes around to show that either number can come first, e.g., 6 + 3 is the same as 3 + 6.

Unit 21 — Addition

Fill in the numbers with all the ways to make 10.

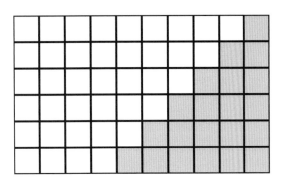 10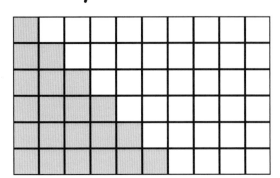

9 + ☐ = 10   ☐ + 9 = 10

☐ + 2 = 10   2 + ☐ = 10

☐ + 3 = 10   3 + ☐ = 10

6 + ☐ = 10   ☐ + 6 = 10

5 + ☐ = 10

Concept: Understand different combinations of number bonds.
Introduction: Use 2 colors of linking cubes. Get the child to link the two colors in different ways to make 10 (1 and 9, 2 and 8, 3 and 7, 4 and 6, 5 and 5). Write the addition equations for the different ways to make 10. Turn each set of linked cubes around to show that either number can come first, e.g., 6 + 4 is the same as 4 + 6.

Unit 22 — Counting On

Kenny collects toy cars. He has 2 cars. Draw 3 more. Count on from 2. How many cars does he have altogether? Write the answer.

2 3 4 5

2 + 3 =

Concept: Count on to add.
Introduction: Set out 6 objects and have the child count them. Write the addition expression 6 + 3. Have the child count on as you set out 3 more objects in a separate set. Write the number he/she counts to for the answer. Get the child to count all the objects to verify that the answer obtained when counting on is correct.

Unit 22 — Counting On

Count the chicks. Draw 4 more chicks. How many chicks does Mother Hen have now? Write the numbers.

5   6   ☐ ☐ ☐

5 + 4 = ☐

Concept: Count on to add.
One Step Further: Tell the child a number between 0 and 15 and ask the child to count on 3 more from that number.

Unit 22 — Counting On

Draw apples on the trees. Count on. Write the answer.

5 + 2 =

6 + 3 =

55

Concept: Count on to add.
Introduction: Write an addition expression, such as 4 + 3. Hold 4 counters. Give the child some counters and ask him/her to hold the second number, 3 counters. Tell him/her that you have 4 counters and he has 3, and ask how many you both have altogether. Get the child to count on from 4 using his/her counters. Repeat, counting on 1, 2, or 3.

Unit 22 — Counting On

Color more blocks. Count on. How many blocks are colored altogether? Write the answer.

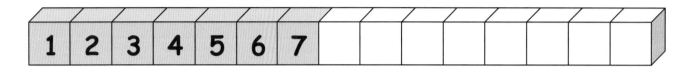

7 + 5 = ☐

9 + 3 = ☐

1 2 3 4 5 6

6 + 6 = ☐

Concept: Count on to add.
Introduction: Write an addition expression, such as 5 + 3. Draw 10 circles on the board in a row. Color the first five and write the numbers in them. Get the child to look at the expression and color 3 more circles and then count on to find the answer.

Unit 22 — Counting On

There are 5 apples in one tree. Draw apples on the other tree. Count on. Write the answer.

5 + 2 = ☐

5 + 5 = ☐

Concept: Count on to add.
Introduction: Put 5 counters in a closed box or bag. Write the expression 5 + 4. Have the child count out 4 counters and put them in a bowl. Tell the child that there are 5 in the box and ask him/her to find how many there are in the box and the bowl altogether.

Unit 22 — Counting On

There are 10 candies in the box. Draw more candies. How many are there altogether? Write the answer.

10 + 2 =

10 + 4 =

10 + 3 =

Concept: Count on to add.

Unit 22 — Counting On

Use two colors. Color the squares to add. Write the answer. The first one is done for you.

| 1 | 2 | 3 | 4 | 5 | 6 | 7 | 8 | 9 | 10 | 11 | 12 | 13 | 14 | 15 | 16 | 17 | 18 | 19 | 20 |

3 + 5 = 8

| 1 | 2 | 3 | 4 | 5 | 6 | 7 | 8 | 9 | 10 | 11 | 12 | 13 | 14 | 15 | 16 | 17 | 18 | 19 | 20 |

6 + 2 =

| 1 | 2 | 3 | 4 | 5 | 6 | 7 | 8 | 9 | 10 | 11 | 12 | 13 | 14 | 15 | 16 | 17 | 18 | 19 | 20 |

7 + 5 =

| 1 | 2 | 3 | 4 | 5 | 6 | 7 | 8 | 9 | 10 | 11 | 12 | 13 | 14 | 15 | 16 | 17 | 18 | 19 | 20 |

9 + 3 =

Concept: Use a number line to add.
Introduction: Have the child put the numeral cards 1-20 in order. Write an addition expression, such as 5 + 3. Have the child point to the 5. Then have the child use his/her finger to hop three times to the next number while he/she counts: 1, 2, 3. Have the child tell you which number he/she lands on. Line up counters, 5 of one color, and 3 of another, under the numbers to verify the answer.

Unit 22 — Counting On

Use two colors. Color the squares to add. Write the answer.

| 1 | 2 | 3 | 4 | 5 | 6 | 7 | 8 | 9 | 10 | 11 | 12 | 13 | 14 | 15 | 16 | 17 | 18 | 19 | 20 |

2 + 7 =

| 1 | 2 | 3 | 4 | 5 | 6 | 7 | 8 | 9 | 10 | 11 | 12 | 13 | 14 | 15 | 16 | 17 | 18 | 19 | 20 |

6 + 9 =

| 1 | 2 | 3 | 4 | 5 | 6 | 7 | 8 | 9 | 10 | 11 | 12 | 13 | 14 | 15 | 16 | 17 | 18 | 19 | 20 |

8 + 5 =

| 1 | 2 | 3 | 4 | 5 | 6 | 7 | 8 | 9 | 10 | 11 | 12 | 13 | 14 | 15 | 16 | 17 | 18 | 19 | 20 |

10 + 6 =

Concept: Use a number line to add.
Introduction: Draw a number line. Write an addition expression, such as 5 + 3. Have the child point to the 5. Then have the child use his/her finger to hop three times to the next number while he/she counts: 1, 2, 3. Have the child tell you which number he/she lands on.

Unit 22 — Counting On

Do the following additions.

| 1 | 2 | 3 | 4 | 5 | 6 | 7 | 8 | 9 | 10 | 11 | 12 | 13 | 14 | 15 | 16 | 17 | 18 | 19 | 20 |

2 + 1 = ☐   1 + 1 = ☐

5 + 2 = ☐   1 + 7 = ☐

2 + 6 = ☐   9 + 2 = ☐

1 + 9 = ☐   2 + 8 = ☐

7 + 0 = ☐   0 + 4 = ☐

Concept: Use a number line to add.
Introduction: Draw a number line on the board. Write some addition expressions and get the child to solve them by putting his finger on the first number, and then hopping forward the correct number of spaces.

Unit 22 — Counting On

Color the bear's quilt according to the following:

2—Red
3—Blue
4—Green
5—Yellow

Concept: Solve simple addition problems.

Unit 22 — Counting On

Do the following additions.

| 1 | 2 | 3 | 4 | 5 | 6 | 7 | 8 | 9 | 10 | 11 | 12 | 13 | 14 | 15 | 16 | 17 | 18 | 19 | 20 |

2 + 3 =   4 + 3 =

5 + 4 =   3 + 7 =

4 + 4 =   9 + 3 =

4 + 9 =   4 + 8 =

6 + 4 =   8 + 3 =

Concept: Use a number line to add.
Introduction: Draw a number line on the board. Write some addition expressions and get the child to solve them by putting his finger on the first number, and then hopping forward the correct number of spaces.

Unit 22 — Counting On

Do the following additions.

| 1 | 2 | 3 | 4 | 5 | 6 | 7 | 8 | 9 | 10 | 11 | 12 | 13 | 14 | 15 | 16 | 17 | 18 | 19 | 20 |

$$\begin{array}{r} 4 \\ +\ 5 \\ \hline \end{array} \qquad \begin{array}{r} 7 \\ +\ 5 \\ \hline \end{array} \qquad \begin{array}{r} 5 \\ +\ 9 \\ \hline \end{array}$$

$$\begin{array}{r} 6 \\ +\ 6 \\ \hline \end{array} \qquad \begin{array}{r} 5 \\ +\ 8 \\ \hline \end{array} \qquad \begin{array}{r} 7 \\ +\ 6 \\ \hline \end{array}$$

$$\begin{array}{r} 5 \\ +\ 6 \\ \hline \end{array} \qquad \begin{array}{r} 6 \\ +\ 8 \\ \hline \end{array} \qquad \begin{array}{r} 9 \\ +\ 6 \\ \hline \end{array}$$

$$\begin{array}{r} 10 \\ +\ 3 \\ \hline \end{array} \qquad \begin{array}{r} 10 \\ +\ 6 \\ \hline \end{array} \qquad \begin{array}{r} 10 \\ +\ 9 \\ \hline \end{array}$$

Concept: Use a number line to add.

Unit 23 — Subtraction

Write the answers.

How many monkeys are in the tree?

How many butterflies are there?

2 monkeys climbed down. How many are left in the tree?

4 − 2 =

3 butterflies flew away. How many are left?

9 − 3 =

Concept: Understand simple subtraction problems.
Introduction: Display 7 objects. Ask the child to count them. Write the number. Take away 4 objects. Ask the child how many are left. Write the subtraction equation 7 – 4 = 3. Tell the child that the " − " symbol means "take away". Read the equation to the child as, "Seven take away four is three."

Unit 23 — Subtraction

How many are left? Write the numbers.

3 - 1 =

5 - 3 =

5 - 2 =

6 - 2 =

Concept: Understand simple subtraction problems.
Introduction: Give the child 8 objects. Ask the child to count them. Write the number down. Then write "– 3" after the 8. Ask the child what the "– 3" means. Take 3 objects away and ask the child how many are left. Write "= 5" after "8 – 3". Ask the child to "read" the final equation. "Eight take away three is five."

Unit 23 — Subtraction

Cross out the correct amount. How many are left? Write the answer.

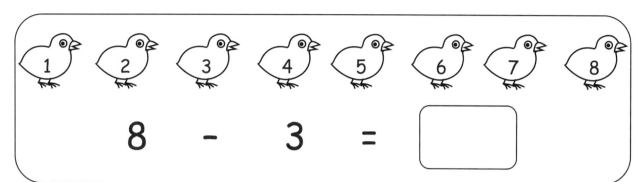

8 - 3 =

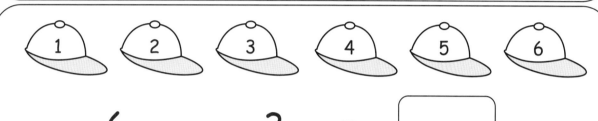

6 - 3 =

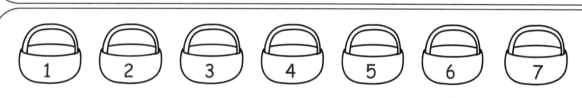

7 - 4 =

10 - 5 =

Concept: Understand simple subtraction problems.
Introduction: Give the child some counters. Ask the child how many will be left if you take away some of the counters. Write the subtraction equation. Tell the child that we can also read "–" as "minus" and "=" as equal. Read the equation, using "minus" and "equal". Tell the child that when we take away some, we 'subtract'. Write some other equations for the child to read.

Unit 23 — Subtraction

Do each subtraction. Write the answer.

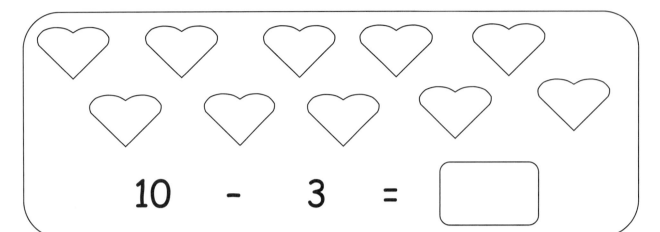

10 - 3 =

7 - 6 =

8 - 8 =

Concept: Understand simple subtraction problems.
Introduction: Give the child some counters. Ask the child how many will be left if you take away some of the counters. Write the subtraction equation. Repeat with other amounts. Include instances where all the counters are taken away, leaving 0.

Unit 23 — Subtraction

Fill in the numbers for each subtraction.

☐ − ☐ = ☐

☐ − ☐ = ☐

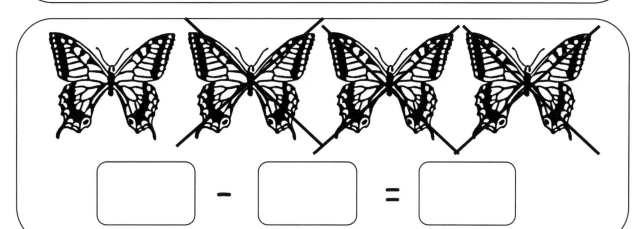

☐ − ☐ = ☐

Concept: Understand simple subtraction problems.
Introduction: Give the child 10 counters. Ask the child how many you need to take away to have 5 left. Write the subtraction equation. Repeat with other amounts.

Unit 23 — Subtraction

Fill in the numbers for each subtraction.

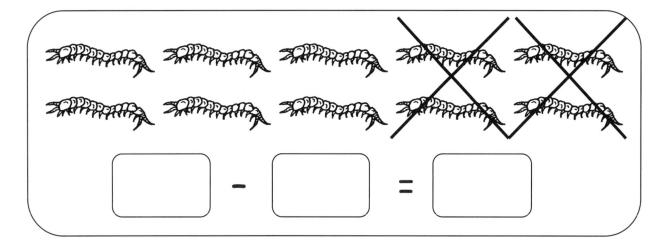

Unit 23 — Subtraction

Draw dots and cross some out to find the answers. Write the answers.

9 - 4 =

5 - 2 =

8 - 4 =

8 - 5 =

7 - 4 =

7 - 6 =

Concept: Understand simple subtraction problems.
One Step Further: Get the child to write his/her own subtraction problems and then work out the answers.

Unit 23 — Subtraction

Draw dots and cross some out to find the answers. Write the answers.

10 − 4 =	10 − 2 =
5 − 4 =	5 − 5 =
6 − 3 =	10 − 7 =

Concept: Understand simple subtraction problems.
Introduction: Write a subtraction expression. Have the child use counters or other objects to find the answer. Write some of the expressions in vertical format. Tell the child that the total goes on top, then the number being taken away, then the answer.

Unit 24 — Part

There are 7 children altogether. 4 of them are boys. How many girls are there? Fill in the numbers.

Concept: Understand subtraction as whole minus a part.
Introduction: Show the child two sets of objects. Ask the child how many there are altogether and draw a number bond. Then, take away one set and ask how many are left. Write the subtraction equation and point to the whole in the number bond, one part, and then the other part as you read the subtraction equation.

Unit 24 — Part

There are 6 sea animal altogether. 4 of them are fishes. The rest are seahorses. How many seahorses are there? Fill in the numbers.

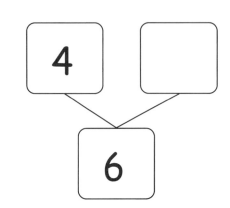

6 - 4 =

There are 8 treats. 5 of them are candies and the rest are lollipops. How many lollipops are there? Fill in the numbers.

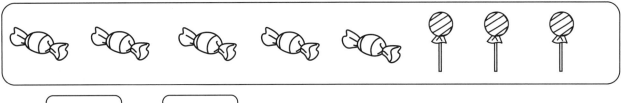

8 - 5 =

Concept: Understand subtraction as whole minus a part.
Introduction: Show the child two sets of objects. Tell the child how many there are altogether. Tell the child each set is a part of the whole. Ask the child to tell you how many are in one of the sets. Write the number bond with an empty part. Then, take away that set and ask the child how many are left. Fill in the remaining number of the number bond, and write the subtraction equation.

Unit 24 — Part

Do each subtraction. Write the answer.

4 - 3 =

10 - 5 =

9 - 3 =

Concept: Understand subtraction as whole minus a part.
Introduction: Show the child two sets of objects. Write a subtraction expression in which one part is subtracted away from the whole. Ask the child to find the answer.

Unit 24 — Part

Do each subtraction. Write the answer.

8 - 4 = ☐

✦ ☾ ✦   ✦ ✦

5 - 1 = ☐

10 - 8 = ☐

Unit 24 — Part

Color. Write the answer.

6 balloons are red. The rest are green. How many green balloons are there?

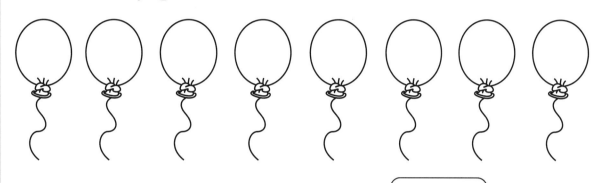

8 - 6 = ☐

2 blocks are blue. The rest are yellow. How many yellow blocks are there?

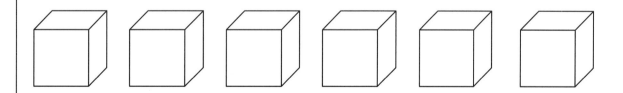

6 - 2 = ☐

Concept: Understand subtraction as whole minus a part.
Introduction: Set out 10 paper cups or other object that have a definite orientation. Tell the child that 3 are upside down. Have the child turn 3 of them over. Ask him/her how many are right side up. Write the equation.

Unit 24 — Part

Color. Write the answer.

7 - 5 =

10 - 1 =

12 - 8 =

Unit 25 — Counting Back

There are 10 balloons. 1 flew away. How many are left?

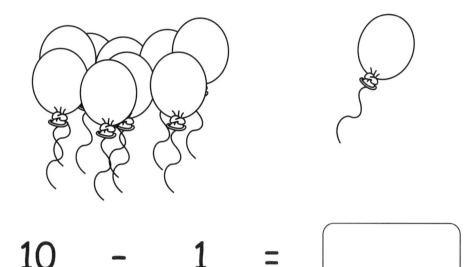

10 - 1 = ☐

There are 6 cupcakes altogether. 2 are not in the box. The rest are in the box. How many are in the box?

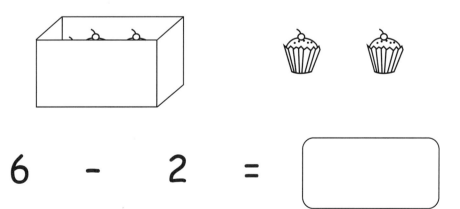

6 - 2 = ☐

Concept: Count back to subtract 1 or 2.
Introduction: Give the child numeral cards 1-10 and have the child put them in order. Say a number between 1 and 10 and ask the child what number is 1 or 2 less. Then remove the cards and repeat, helping the child count back. Write the expression 9 – 1. Show the child a handful of 9 counters. Take away 1 counter and ask how many are left. Get the child to give you the answer without counting the remaining counters. He/she can then count to verify.

Unit 25 — Counting Back

Color the picture according to the following:

2—Blue
3—Green
4—Brown
5—Yellow

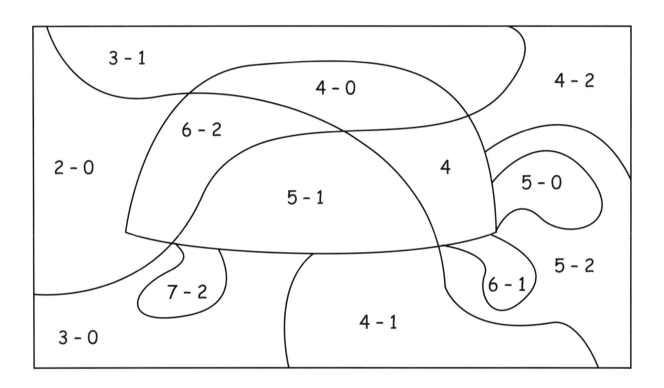

Concept: Count back to subtract 1 or 2.
One Step Further: Say a number within 10 and ask the child to count back 2 from it. Repeat with other numbers within 20.

Unit 25 — Counting Back

Cross out blocks to find the answer. The first one has been done for you. Write the answer.

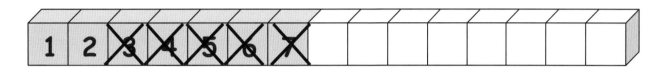

7 - 5 =

9 - 6 =

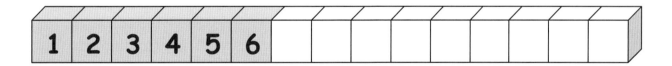

6 - 4 =

Concept: Use a number line to subtract.
Introduction: Write a subtraction expression such as 9 – 5. Give the child the numeral cards for 1-9 and have him/her put them in order. Line up counters underneath the cards. Ask the child how many counters there are. Guide the child in removing 5 counters and 5 cards, starting with the 9 and going backwards. Ask the child how many counters are left, and what the highest number on the cards that are left. Finish the equation: 9 – 5 = 4. Repeat with other examples.

Unit 25 — Counting Back

Color squares and cross some out to find the answer to the subtraction. Write the answer.

| 1 | 2 | 3 | 4 | 5 | 6 | 7 | 8 | 9 | 10 | 11 | 12 | 13 | 14 | 15 | 16 | 17 | 18 | 19 | 20 |

9 - 5 = ☐

| 1 | 2 | 3 | 4 | 5 | 6 | 7 | 8 | 9 | 10 | 11 | 12 | 13 | 14 | 15 | 16 | 17 | 18 | 19 | 20 |

13 - 2 = ☐

| 1 | 2 | 3 | 4 | 5 | 6 | 7 | 8 | 9 | 10 | 11 | 12 | 13 | 14 | 15 | 16 | 17 | 18 | 19 | 20 |

20 - 5 = ☐

| 1 | 2 | 3 | 4 | 5 | 6 | 7 | 8 | 9 | 10 | 11 | 12 | 13 | 14 | 15 | 16 | 17 | 18 | 19 | 20 |

14 - 6 = ☐

Concept: Use a number line to subtract.
Introduction: Have the child put the numeral cards 1-20 in order. Write a subtraction expression, such as 14 –5. Ask the child to point to the 14. Then ask the child to use his/her finger to hop three five times backwards while he/she counts to 5. Ask the child which number he/she lands on. Line up counters under the cards, and remove
5 from the end to verify the answer. Finish the equation: 14 – 5 = 9

Unit 25 Counting Back

Do the following subtractions.

| 1 | 2 | 3 | 4 | 5 | 6 | 7 | 8 | 9 | 10 | 11 | 12 | 13 | 14 | 15 | 16 | 17 | 18 | 19 | 20 |

11 − 3 = ☐   10 − 6 = ☐

8 − 5 = ☐   12 − 7 = ☐

13 − 6 = ☐   19 − 3 = ☐

17 − 7 = ☐   12 − 8 = ☐

20 − 0 = ☐   20 − 4 = ☐

Concept: Use a number line to subtract.
Introduction: Draw a number line on the board. Write some subtraction expressions and get the child to solve them by putting his finger on the first number, and then hopping backward the correct number of spaces.

Unit 25 — Counting Back

Do the following subtractions.

1	2	3	4	5	6	7	8	9	10	11	12	13	14	15	16	17	18	19	20

$$14 - 5 =$$

$$7 - 5 =$$

$$15 - 9 =$$

$$16 - 6 =$$

$$15 - 8 =$$

$$7 - 7 =$$

$$17 - 6 =$$

$$16 - 4 =$$

$$19 - 9 =$$

$$20 - 3 =$$

$$20 - 6 =$$

$$20 - 11 =$$

Concept: Use a number line to subtract.
Introduction: Draw a number line on the board. Write some subtraction expressions and get the child to solve them by putting his finger on the first number, and then hopping backward the correct number of spaces.

Unit 26 — Addition and Subtraction

Count on to add. Cross off to subtract. Write the answers.

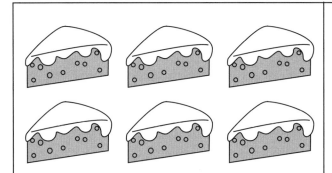

6 + 2 =

6 − 2 =

9 + 1 =

9 − 1 =

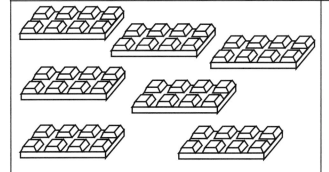

7 + 2 =

7 − 2 =

4 + 1 =

4 − 1 =

Concept: Add or subtract from a number within 10.
Introduction: Give the child 5 objects. Ask him/her how many there will be if 2 are added. Write the equation. Ask him/her how many there will be if 2 are taken away. Write the equations.

Unit 26 Addition and Subtraction

Color more to add. Cross out to subtract. Write the answers.

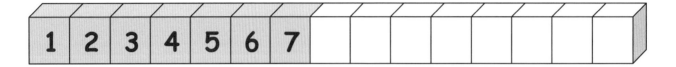

7 + 4 = ☐   7 − 4 = ☐

9 + 5 = ☐   9 − 5 = ☐

6 + 6 = ☐   6 − 6 = ☐

Concept: Use a number line to subtract.
Introduction: Draw a number line on the board. Write two equations, adding and subtracting from the same number. Guide the child in using the number line to find the answer.

Unit 26 — Addition and Subtraction

Write the answers.

4 + 3 = ☐ | 4 , 3 → 7 | 7 − 3 = ☐

3 + 4 = ☐ | 7 − 4 = ☐

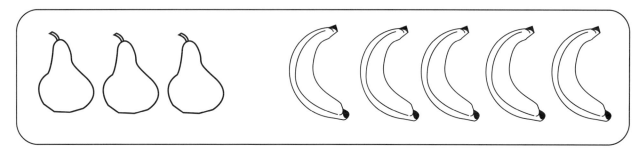

5 + 3 = ☐ | 3 , 5 → 8 | 8 − 3 = ☐

3 + 5 = ☐ | 8 − 5 = ☐

Concept: Write a family of addition and subtraction equations.
Introduction: Show the child two sets of objects, such as 3 pencils and 2 pens. Write a number bond with empty circles and get the child to fill it in. Then ask questions based on the situation leading to 2 addition equations and 2 subtraction equations, such as, "There are 3 pencils and 2 pens. How many things to write with are there altogether? There are 2 pens and 3 pencils. How many things to write with are there altogether? There are 5 pens and pencils. 3 of them are pencils. How many pens are there? 2 of them are pens. How many pencils are there?" Write the equations.

Unit 26 — Addition and Subtraction

Write the answers.

7 + 3 =
3 + 7 =

10 − 3 =
10 − 7 =

6 + 2 =
2 + 6 =

8 − 2 =
8 − 6 =

5 + 4 =
4 + 5 =

9 − 4 =
9 − 5 =

Concept: Write a family of addition and subtraction equations.

Unit 26 — Addition and Subtraction

Do the following additions and subtractions.

| 1 | 2 | 3 | 4 | 5 | 6 | 7 | 8 | 9 | 10 | 11 | 12 | 13 | 14 | 15 | 16 | 17 | 18 | 19 | 20 |

2 + 2 = ☐   3 − 2 = ☐

4 + 3 = ☐   8 − 4 = ☐

8 + 5 = ☐   12 − 6 = ☐

4 + 2 = ☐   6 − 3 = ☐

6 + 7 = ☐   15 − 5 = ☐

Concept: Use a number line to add or subtract.
Introduction: Draw a number line on the board. Write some addition and subtraction expressions and get the child to solve them using the number line.

Unit 26 — Addition and Subtraction

Do the following additions and subtractions.

| 1 | 2 | 3 | 4 | 5 | 6 | 7 | 8 | 9 | 10 | 11 | 12 | 13 | 14 | 15 | 16 | 17 | 18 | 19 | 20 |

$$\begin{array}{r} 4 \\ +\ 5 \\ \hline \end{array}$$

$$\begin{array}{r} 7 \\ +\ 5 \\ \hline \end{array}$$

$$\begin{array}{r} 7 \\ -\ 3 \\ \hline \end{array}$$

$$\begin{array}{r} 10 \\ +\ 6 \\ \hline \end{array}$$

$$\begin{array}{r} 11 \\ +\ 3 \\ \hline \end{array}$$

$$\begin{array}{r} 17 \\ -\ 7 \\ \hline \end{array}$$

$$\begin{array}{r} 7 \\ +\ 4 \\ \hline \end{array}$$

$$\begin{array}{r} 6 \\ -\ 4 \\ \hline \end{array}$$

$$\begin{array}{r} 8 \\ -\ 3 \\ \hline \end{array}$$

$$\begin{array}{r} 9 \\ +\ 2 \\ \hline \end{array}$$

$$\begin{array}{r} 9 \\ -\ 6 \\ \hline \end{array}$$

$$\begin{array}{r} 20 \\ -10 \\ \hline \end{array}$$

Concept: Use a number line to subtract.
Introduction: Draw a number line on the board. Write some addition and subtraction expressions and get the child to solve them using the number line.

Unit 26 — Addition and Subtraction

Read the story and answer the question. Fill in the blank.

Sally has 3 apples.

Her mother gives her 2 more apples.

How many apples does Sally have now?

Sally has _____ apples now.

Concept: Understand simple story problems.
Introduction: Make up a simple story problems and have the child act it out with objects to solve.
Using This Page: Read the story to the child.
One Step Further: Get the child to write an equation for this problem.

Unit 26 — Addition and Subtraction

Read the story and answer the question. Fill in the blank.

John has 8 toy trucks.

One day, he lost 2 of them.

How many toy trucks does John have now?

John has _____ toy trucks now.

Unit 26 — Addition and Subtraction

Read the story and answer the question. Fill in the blank.

Sam has 5 stamps.

Mark has 3 stamps.

How many stamps do Sam and Mark have altogether?

Sam and Mark have _____ stamps altogether.

Unit 26 — Addition and Subtraction

Read the story and answer the question. Fill in the blank.

Maria has 4 teddy bears.

She gives 1 to her sister.

How many teddy bears does Maria have now?

Maria has _____ teddy bears now.

Concept: Understand simple story problems.
Introduction: Make up a simple story problems and have the child act it out with objects to solve.
Using This Page: Read the story to the child.
One Step Further: Get the child to write an equation for this problem.

Unit 26 — Addition and Subtraction

Read the story and answer the question. Fill in the blank.

Wendy has 4 fishes.

Her father gives her 4 more fishes.

How many fishes does Wendy have now?

Wendy has _____ fishes now.

Concept: Understand simple story problems.
Introduction: Make up a simple story problems and have the child act it out with objects to solve.
Using This Page: Read the story to the child.
One Step Further: Get the child to write an equation for this problem.

Unit 26 — Addition and Subtraction

Read the story and answer the question. Fill in the blank.

There are 5 cakes on the table.

The cat eats 3 of them.

How many cakes are there now?

There are _____ cakes left.

Unit 26 — Addition and Subtraction

Color the shape in each row that shows a **greater** number value.

1	2	3	4	5	6	7	8	9	10	11	12	13	14	15	16	17	18	19	20

1 + 1	0 + 3
2 + 2	1 + 2
4 + 1	3 + 1
2 + 3	5 + 2

Concept: Practice addition and subtraction, compare numbers.
Introduction: Write a 4 and a 5 on the board and ask the child which number is greater. Under the 4 write the addition expression 2 + 2 and under the 5 write the addition expression 2 + 3. Allow the child to use a number line, objects, or count on to find the answers. Then ask the child which answer is greater. Write two other simple addition expressions and get the child to find the answer and tell you which answer is greater.

Unit 26 — Addition and Subtraction

Color the shape in each row that shows a **greater** number value.

| 1 | 2 | 3 | 4 | 5 | 6 | 7 | 8 | 9 | 10 | 11 | 12 | 13 | 14 | 15 | 16 | 17 | 18 | 19 | 20 |

0 + 7	0 + 9
2 + 3	10 − 7
7 − 3	4 + 2
9 + 3	9 − 3

98

Concept: Practice addition and subtraction, compare numbers.
Introduction: Write two addition or subtraction expressions on the board. Ask the child to find the answers, and then tell you which answer is greater.

Unit 26 — Addition and Subtraction

Color the shape in each row that shows a **smaller** number value.

| 1 | 2 | 3 | 4 | 5 | 6 | 7 | 8 | 9 | 10 | 11 | 12 | 13 | 14 | 15 | 16 | 17 | 18 | 19 | 20 |

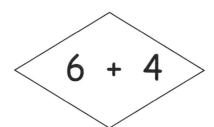

6 + 4   8 + 4

7 − 2   10 + 2

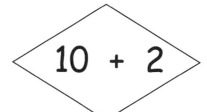

1 + 8   6 − 3

3 + 5   9 + 1

Concept: Practice addition and subtraction, compare numbers.
Introduction: Write two numbers and ask the child which is smaller. Write two addition or subtraction expressions and ask the child to find the answer. Then ask him/her which answer is smaller.

Unit 26 — Addition and Subtraction

Color the shape in each row that shows a **smaller** number value.

| 1 | 2 | 3 | 4 | 5 | 6 | 7 | 8 | 9 | 10 | 11 | 12 | 13 | 14 | 15 | 16 | 17 | 18 | 19 | 20 |

10 + 4   16 − 6

8 − 2   8 − 3

5 + 5   7 + 4

14 − 3   6 + 6

Concept: Practice addition and subtraction, compare numbers.
Introduction: Write two numbers and ask the child which is smaller. Write two addition or subtraction expressions and ask the child to find the answer. Then ask him/her which answer is smaller.

Unit 27 — Numbers to 40

Fill in the blanks.

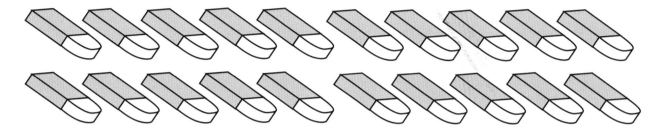

There are _____ groups of tens.

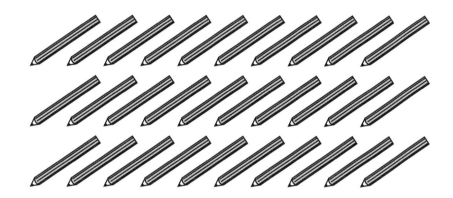

There are _____ groups of tens.

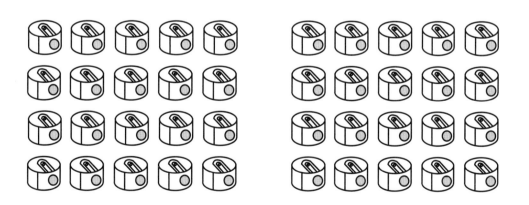

There are _____ groups of tens.

Concept: Count groups of tens.
Introduction: Give the child 40 counters and ask him/her to make groups of ten. Then ask the child how many groups of 10 he/she has.

Unit 27 — Numbers to 40

Count and write the correct number.

10	🍓🍓🍓🍓🍓🍓🍓🍓🍓🍓
	🍓🍓🍓🍓🍓🍓🍓🍓🍓🍓 🍓🍓🍓🍓🍓🍓🍓🍓🍓🍓
	🍓🍓🍓🍓🍓🍓🍓🍓🍓🍓 🍓🍓🍓🍓🍓🍓🍓🍓🍓🍓 🍓🍓🍓🍓🍓🍓🍓🍓🍓🍓
	🍓🍓🍓🍓🍓🍓🍓🍓🍓🍓 🍓🍓🍓🍓🍓🍓🍓🍓🍓🍓 🍓🍓🍓🍓🍓🍓🍓🍓🍓🍓 🍓🍓🍓🍓🍓🍓🍓🍓🍓🍓

Concept: Count groups of tens.
Introduction: Use linking cubes or other objects grouped in ten. Give the child 1 ten and ask how many tens there are. Write the number 10. Repeat with 2 tens and 20, 3 tens and 30, 4 tens and 40. Tell the child that 30 is "thirty" and 40 is "forty". Point to the numbers in the tens place and explain that the number in this place is the number of groups of tens. Point to the numbers in the ones place and ask the child what the 0 means. Elicit the response that this means there are no ones.

Unit 27 — Numbers to 40

Count the tens and ones. Write the tens and ones and the number.

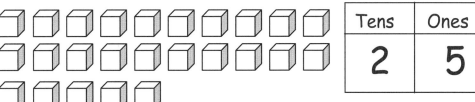

Tens	Ones
2	5

25

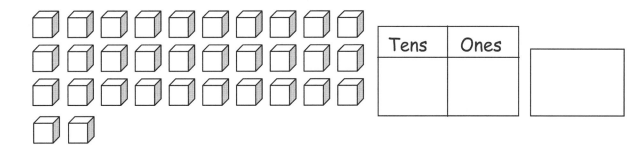

Tens	Ones

Tens	Ones

Tens	Ones

Tens	Ones

Concept: Count to 40 in tens and ones.
Introduction: Give the child between 30 and 40 objects and have him/her make groups of tens, count the groups of tens, and count the leftovers. Write the number as tens and ones and get the child to read it.
One Step Further: Get the child to read the numbers he/she has written on this page, e.g., "thirty-two".

Unit 27 — Numbers to 40

Count the tens and ones. Write the number of dots in each row.

Unit 27 — Numbers to 40

Circle groups of tens. Write how many shells and how many flowers there are.

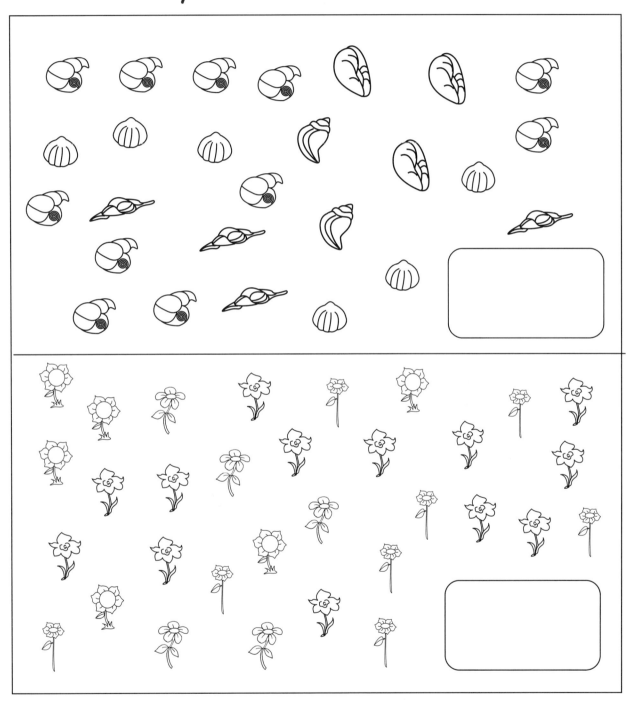

Unit 27 — Numbers to 40

Color the correct number of squares.

25

33

27

106

Concept: Count to 40 in tens and ones.
Introduction: Use linking cubes in groups of tens along with individual cubes. Write a number between 20 and 40 and have the child show you that number with objects.

Unit 27 — Numbers to 40

Read the numbers.

1	2	3	4	5
6	7	8	9	10
11	12	13	14	15
16	17	18	19	20
21	22	23	24	25
26	27	28	29	30
31	32	33	34	35
36	37	38	39	40

Fill in the missing numbers.

21	22			25
	27	28		30
31			34	35
36		38		40

Concept: Count to 40.
Introduction: Use the chart on this page and get the child to count to 40. Guide the child in reading the rows in order, going back to the beginning of the row for each new row. Have the child count out 40 counters.

Unit 27 — Numbers to 40

Match the objects with the numbers.

20 •   •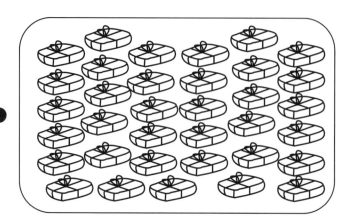

25 •   •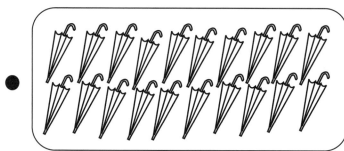

30 •   •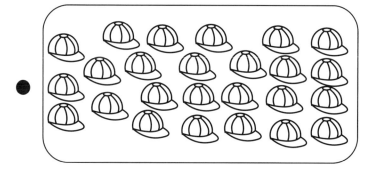

35 •   •

Concept: Count to 40.
Introduction: Write a number between 21 and 40 and have the child count out the correct number of counters.

Unit 27 — Numbers to 40

Connect the numbers in order.

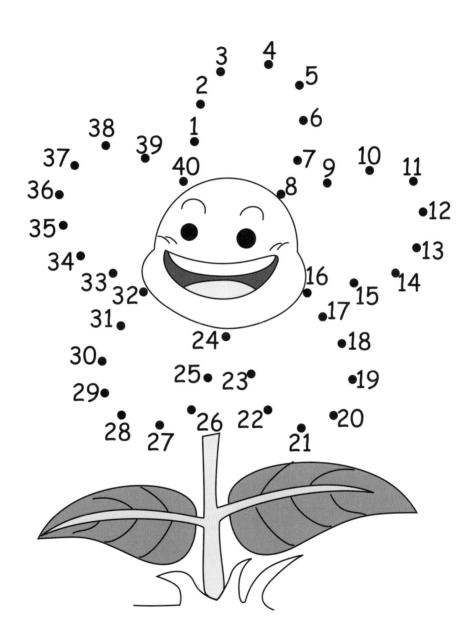

Concept: Numbers have order.
Introduction: Give the child the numeral cards 21-40 and have him/her put them in order.

Unit 27 — Numbers to 40

Fill in the missing numbers.

16 17 ☐ ☐ 20

21 ☐ 23 ☐ ☐

26 ☐ ☐ 29 ☐

31 ☐ ☐ ☐ 35

36 37 ☐ ☐ ☐

Concept: Write numbers in order from 1 to 40.

Unit 27 — Numbers to 40

Connect the numbers in order. What do you see?

Concept: Numbers have order.
Introduction: Give the child a set of 5 consecutive numeral cards for numbers between 15 and 40 and have him/her put them in order.

Unit 27 — Numbers to 40

Fill in the missing numbers.

Unit 27 — Numbers to 40

Connect the numbers in order. What do you see?

Concept: Numbers have order.
Introduction: Say a number between 2 and 40 and ask the child what number comes after and what number comes before. Repeat with other numbers.

Unit 27 — Numbers to 40

How many lollipops are there? Write the number in the lollipops as you count them.

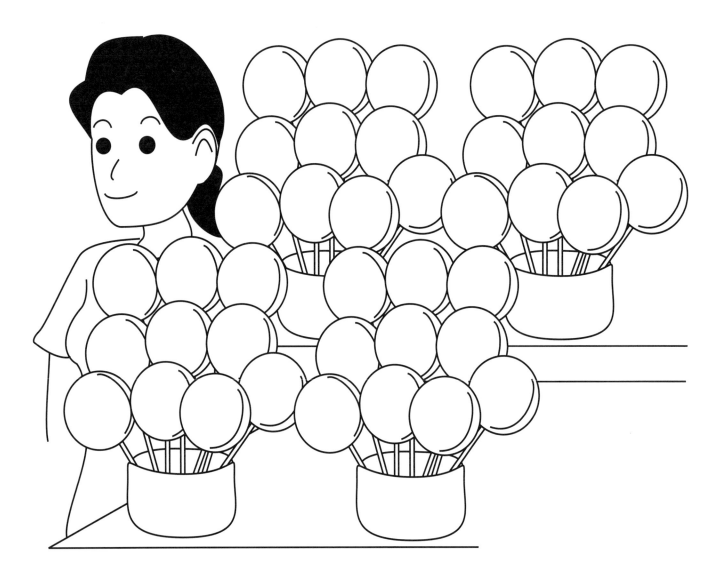

Concept: Numbers have order.
Introduction: Give the child numeral cards 1-40 and have him/her put them in order. Ask the child to tell you the number that come before 20, before 30, and before 40.

Unit 27 — Numbers to 40

Connect the numbers in order. What do you see?

Concept: Numbers have order.

Unit 27 — Numbers to 40

Fill in the missing numbers.

21 22 ☐ ☐ 25

18 ☐ 20 ☐ ☐

28 ☐ ☐ 31 ☐

25 ☐ ☐ ☐ 29

36 37 ☐ ☐ ☐

Concept: Numbers have order.
Introduction: Say a number and ask the child to count up from that number. Have numeral cards 1-40 available. Select 2 non-consecutive cards and get the child to select the cards that come in between them and put them in order.

Unit 27 — Numbers to 40

Start at 40 and connect the numbers in count-down order. What do you see?

Concept: Numbers have reverse order.
Introduction: Give the child the number cards 1-40 and have him/her put them in reverse order. Ask the child to count backwards two numbers from a specific number, for example, point to 31 and have the child count 30, 29.

Unit 27 — Numbers to 40

Fill in the missing numbers.

[   ] 39 [   ] 37 36

[   ] [   ] 33 32 [   ]

[   ] 29 [   ] 27 [   ]

[   ] [   ] [   ] 22 21

[   ] 19 [   ] [   ] 16

Concept: Write numbers to 40.
Introduction: Help the child count backwards from 40. Use a number chart to help.

Unit 28 — Ordering

Look at the pictures below. Tell a story in your own words. What happens in the end?

1.

2.

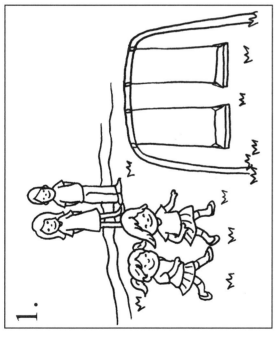

3.

4.

Concept: Understand the order of events.
Introduction: Tell the child a simple story. Afterwards, ask questions about the sequence, "What happened in the beginning? What happened next? After ____ then what happened?".
Using This Page: Point out to the child that the numbers 1, 2, 3, and 4 are used to tell them the order of the story. Ask them what number is used for the beginning, what number is used for what happens next, and so on. The story pictures are in the same order as the numbers on the pictures.
One Step Further: Have the child illustrate a story through four pictures in sequence.

Unit 28 — Ordering

Write numbers to show the correct order.

1

120

Concept: Understand the order of events.
Introduction: Ask the child to describe the events of his/her morning. Then ask the child questions involving the order. For example: "Did you eat breakfast before you got up?"
Using This page: Have the child look at the pictures and say which should come first. Point out the 1 written in the box in the corner of the picture. Ask him/her which picture should come next and tell them to write a 2 in that box. Continue guiding the child in numbering the pictures.

Unit 28 — Ordering

Look at each set of pictures. Circle the picture of what came first.

Unit 28 — Ordering

Write the numbers 1, 2, and 3 to show the correct order.

Concept: Understand the order of events and the concept of first.
Introduction: Have pictures that show a sequence of events. Mix them up and ask the children to arrange them in correct order. Use the term 'first' for what happened first.
One Step Further: Ask the child to describe or explain the order they chose for the pictures on this page. Encourage him/her to use the term 'first', followed by 'then' or 'next'.

Unit 28 — Ordering

Write the numbers 1, 2, and 3 to show the correct order.

Concept: Understand the order of events.
Introduction: Have the child to describe different stages of growth in order, for themselves or for something they are learning in some other area of the curriculum, such as science.
One Step Further: Ask the child to describe or explain the order he/she chose for the pictures on this page.

Unit 28 — Ordering

Write the numbers 1, 2, and 3 to show the correct order.

Concept: Understand the order of events.
Introduction: Have the child to describe the steps needed for a specific task or activity in order, such as the order in which he/she gets dressed.
One Step Further: Ask the child to describe or explain the order he/she chose for the pictures on this page or to make up a story to go along with the pictures.

Unit 28 — Ordering

Write the numbers 1, 2, and 3 to show the correct order.

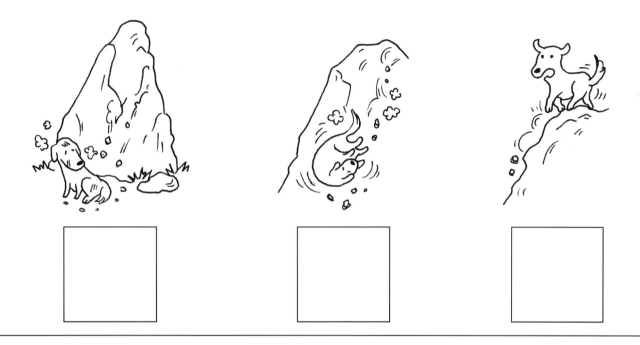

Unit 28 — Ordering

Write the numbers 1, 2, and 3 to show the correct order.

126

Concept: Understand the order of events.
Introduction: Ask the child to give you directions for some activity. Follow his/her directions.
One Step Further: Ask the child to describe or explain the order he/she chose for the pictures on this page or to make up a story to go along with the pictures.

Unit 28 — Ordering

Color the sets that are arranged in order by size.

Concept: Understand arrangement by size.
Introduction: Give the child four objects of different sizes and ask him/her to arrange them by size. Then ask which object is first if they are arranged from smallest to biggest, and which one is first if they are arranged from biggest to smallest.

Unit 28 — Ordering

Circle the sets of numbers that are arranged in order from smallest to greatest.

2	7	10	13
6	4	9	12
1	5	10	8
5	9	8	20

Concept: Order numbers.
Introduction: Give the child 4 non-consecutive numeral cards between 1 and 20. Provide linking cubes and have the child make towers with the corresponding number of cubes. Ask the child to put the towers in order from smallest to largest. Arrange the number cards in the same order. Tell the child that the numbers are in order from smallest to greatest.

Unit 28 — Ordering

Rewrite the numbers in order from the smallest to the greatest.

Concept: Order numbers.
Introduction: Have the child put the numeral cards 1-20 in order. Pull out 4 non-consecutive numbers, mix them up, and have the child put them in order in a separate row. Remove all the number cards, and give the child 4 other numbers to put in order. Demonstrate the following procedure: Count up until you get to one of the numbers. That one will be the smallest. Continue counting until you get to the next number, which will be next, and so on.

Unit 28 — Ordering

Rewrite the numbers in order from the smallest to the greatest.

9   12   10   14

7   17   15   20

18   16   11   13

Concept: Order numbers.
Introduction: Give the child 3-4 numeral cards between 1 and 40 to arrange in order.

Unit 28 — Ordering

Color the child who is first in line.

Concept: Understand first in line.
Introduction: Line up different objects and ask the child to point to the first object. Line them up again differently, and ask the child to point to the first in the new line.

Unit 28 — Ordering

Match each child with his position.

Finishing line

| first
1st | second
2nd | third
3rd | fourth
4th | fifth
5th |

Concept: Understand ordinal number.
Introduction: Line up a row of objects. Point to each one in order and say its ordinal position. Then ask the child to point to the first, second, etc.
Using This Page: Read the words for the ordinal numbers. Explain that we can write the ordinal number with a number and part of a word, e.g., 1st. Get the child to read the ordinal numbers.

Unit 28 — Ordering

Circle the child in the given position.

(School bus with line of 5 children)	3rd
(Shop with line of 5 children)	2nd
(Library with line of 5 children)	5th
(Bus stop with line of 6 children)	1st

133

Concept: Understand ordinal number.
Introduction: Line up a row of objects. Let the child call out the ordinal number while pointing to the object.

Unit 28 — Ordering

Circle the object in the given position.

4th	
1st	
2nd	
3rd	
5th	

Concept: Understand ordinal number.
Introduction: Line up a row of objects. Call out an ordinal number and have the child point to the object in the correct position.

Unit 29 — Time

Color the picture. Is it day or night?

Concept: Understand ways to measure time.
Introduction: Talk with the child about the differences between day and night. Tell the child that we use 'day' and 'night' to talk about different times.

Unit 29 — Time

Where should the girl be? Draw a line to the correct picture.

Concept: Understand ways to measure time.
Introduction: Talk with the child about the different things he/she does during the morning, afternoon, or evening. Tell the child that we use 'morning', 'afternoon', and 'evening' to divide up the day into different times.

Unit 29 — Time

Draw a line to match the time to the clock.

4 o'clock

9 o'clock

12 o'clock

Concept: Understand ways to measure time.
Introduction: Tell the child that we divide the day up into hours. Talk about activities that take about an hour. Show the child an analog clock. Tell him/her there are 12 hours from the middle of the night to the middle of the day, and 12 hours from the middle of the day to the middle of the night. Ask the child to read the numbers on the clock. Draw the child's attention to the hands. Explain that the hour starts when the long hand points to 12 and the short hand points to the hour. Move the hands to demonstrate how they move. Set the time to 3:00 and say that this is 3 o'clock. Set the time at various hours and get the child to say what time it is.

Unit 29 — Time

Write the correct time.

____:00

____:00

____:00

____:00

____:00

____:00

Concept: Understand ways to measure time.
Introduction: Discuss what time the child does various activities throughout the day. For example, the child might get up at 7 o'clock. Show the child a digital clock. Tell him/her that this is another way of showing what time it is. The first number before the two dots is the hour. When the two numbers after the dots is '00' then it is the start of the hour, or o'clock.
One Step Further: Draw the child's attention to the clock, both analog and digital, at various times during the day.

Unit 29 — Time

Fill in the missing numbers.

April

Sunday	Monday	Tuesday	Wednesday	Thursday	Friday	Saturday
	1	2	3	4		
7	8	9				13
14				18	19	
	22	23	24	25		
		30				

Concept: Understand ways to measure time.
Introduction: Discuss yearly activities, such as birthdays. Tell the child that we divide up the year into 12 months. Name the months and discuss things that happen each month. Tell the child we divide the year into weeks and that a week has 7 days. Name the days of the week and talk about what the child does on the various days of the week. Show the child a calendar and the different pages for each month and the rows for each week. Discuss things the child did yesterday, is doing today, and will do tomorrow, pointing to the days on the calendar and using the name of week day, e.g., "Yesterday, on Tuesday, you…"

Unit 29 — Time

Fill in the missing numbers. Color all the days that are Sunday.

May

Sunday	Monday	Tuesday	Wednesday	Thursday	Friday	Saturday
			1			4
5	6		8	9	10	
12		14				18
	20	21			24	25
				30	31	

Concept: Understand ways to measure time.
Introduction: Show the child a calendar. Look at the different months and discuss how many days are in each month. Show the child how each month starts on the next day, so that if one month ends on a Thursday, the next month starts on the next day, a Friday. Ask the child to point to the days in a month that are on a particular week day. They are all in the same column.
One Step Further: Make a large calendar page for the current month. Each day, discuss what day of the week it is. Get the child to illustrate the calendar with special events of the day, or the weather.

Unit 30 — Numbers to 100

Write the number of tens.

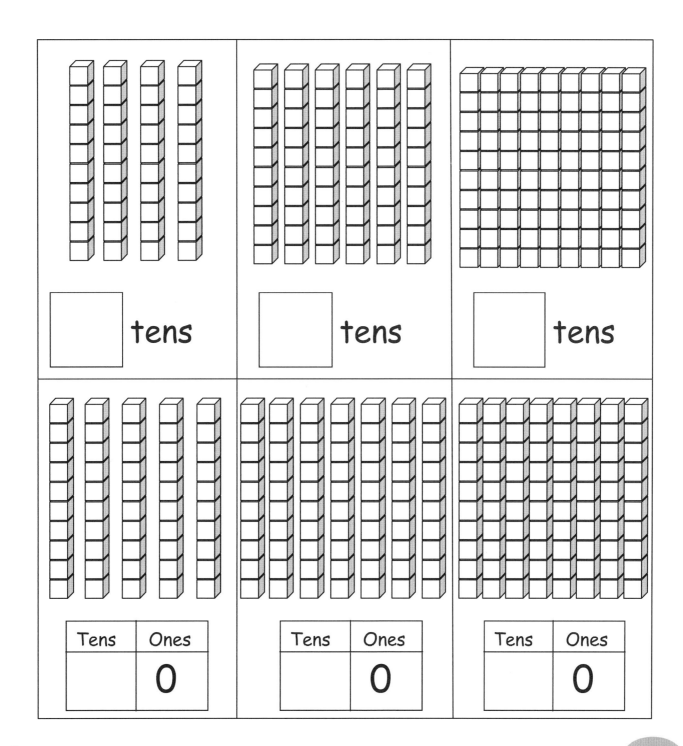

Concept: Count groups of tens.
Introduction: Give the child 100 counters and ask him/her to make groups of ten. Then ask the child how many groups of 10 he/she has.

Unit 30 — Numbers to 100

Count and write the correct number in each row.

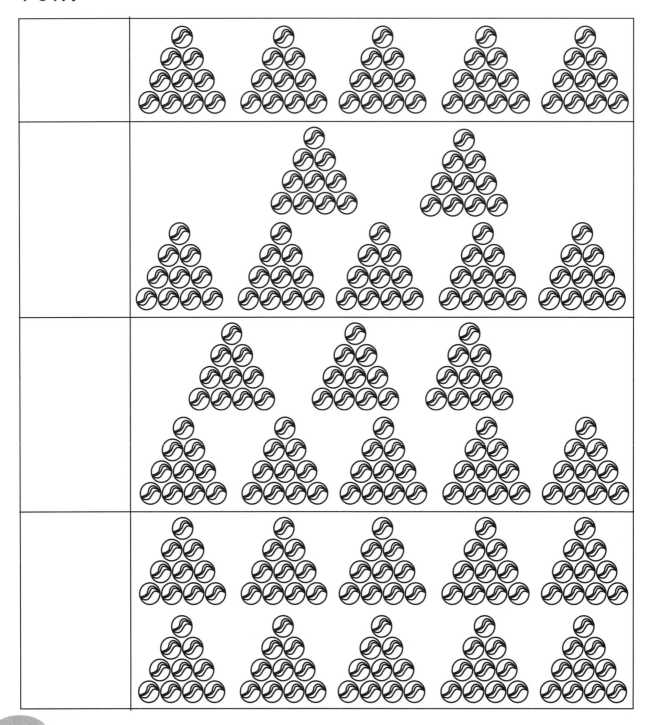

Concept: Count groups of tens.
Introduction: Use linking cubes or other objects grouped in ten. Give the child 4 tens and ask how many tens there are. Write the number 40. Repeat with tens up to 100. Read the numbers to the child. Get the child to count out loud by tens.

Unit 30 — Numbers to 100

Count the tens and ones. Write the tens and ones and the number.

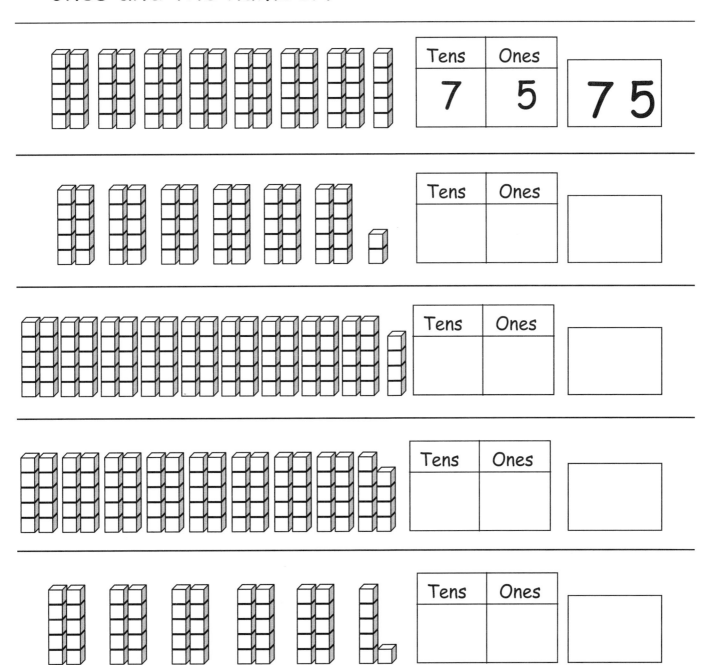

Unit 30 — Numbers to 100

Fill in the missing numbers. Read the numbers.

1	2	3		5	6	7	8		10
11		13	14	15	16	17			20
	22	23	24		26	27	28	29	
31	32	33		35	36		38	39	40
	42		44	45	46	47	48		50
51		53	54	55		57	58	59	60
61	62	63	64		66	67		69	
	72	73		75	76	77	78	79	80
81	82		84	85		87	88		90
91	92	93	94		96	97	98	99	100

Concept: Count to 100.
Introduction: Get the child to count out loud to 100.
One Step Further: Get the child to look for patterns in the hundreds chart.

Unit 30 — Numbers to 100

Count by 5's. Write the numbers.

1	2	3	4		6	7	8	9	
11	12	13	14		16	17	18	19	
21	22	23	24		26	27	28	29	
31	32	33	34		36	37	38	39	
41	42	43	44		46	47	48	49	

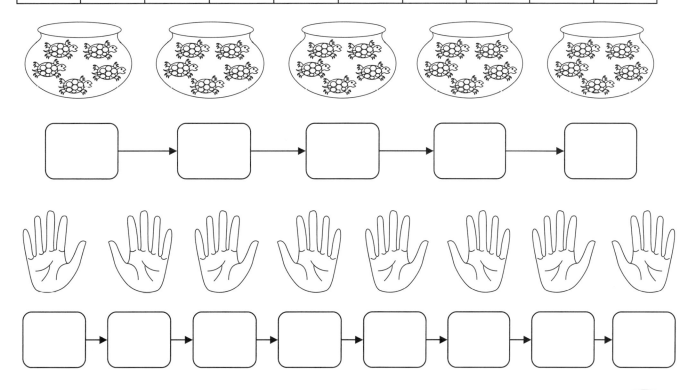

Concept: Count by fives.
Introduction: Use linking cubes linked in 5's. Set out 5 and ask the child how many there are. Then set out another 5 and get the child to count to himself/herself up 5 and tell you the answer out loud. Continue for ten 5's. Point to each 5 and count out loud by 5's.
One Step Further: Have the child practice counting by 5's out loud.

Unit 30 — Numbers to 100

Count by 2's. Write the numbers.

1		3		5		7		9	
11		13		15		17		19	

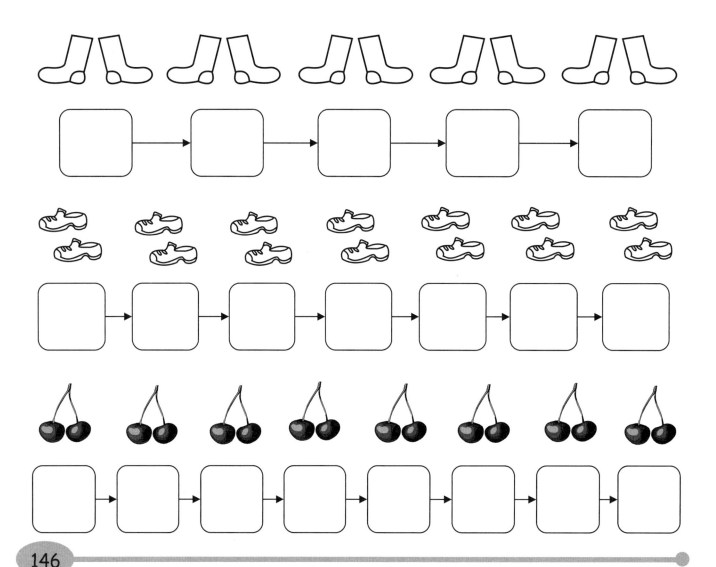

Concept: Count by twos.
Introduction: Set out 20 pair of objects. Move aside one pair of objects and get the child to count them. Move another pair of objects over and get the child to tell you how many there are by counting up 2. Continue to 20. Point to the pairs and count by 2's.
One Step Further: Have the child practice counting by 2's out loud.

Unit 31 — Even/Odd

Draw lines to connect pairs of toys.

There are ☐ pairs of toys.

There are ☐ toys in all.

Concept: Grouping by twos.
Introduction: Provide five pairs of socks or other items that come in matched pairs. Ask the child to match them by pairing them together. Tell the child that each set is a 'pair'. Ask them how many pairs of socks there are, and how many total socks there are in all. Give the child an odd set of socks and get the child to make pairs. Tell him/her that the extra object is the 'odd one out'.

Unit 31 — Even/Odd

Circle sets that have an even number.

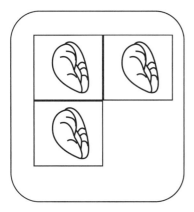
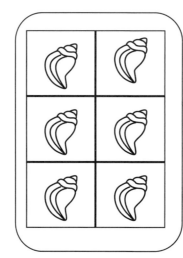
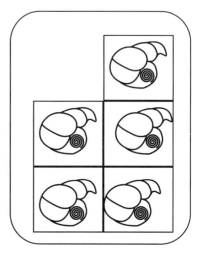

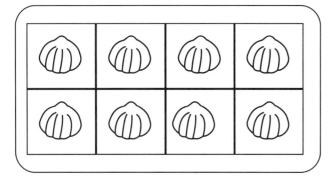
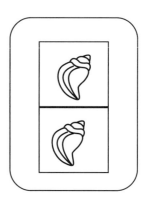

Concept: Understand even and odd numbers.
Introduction: Display 10 blocks. Ask the child to put blocks in towers of two. Tell him/her that 10 is an even number because they can put 10 objects in pairs with nothing left over. Give the child up to 10 linking cubes and have him/her link them in pairs. Ask how many he/she has, and if it is an even number.

Unit 31 — Even/Odd

Circle pairs in each box. Cross out boxes with odd numbers.

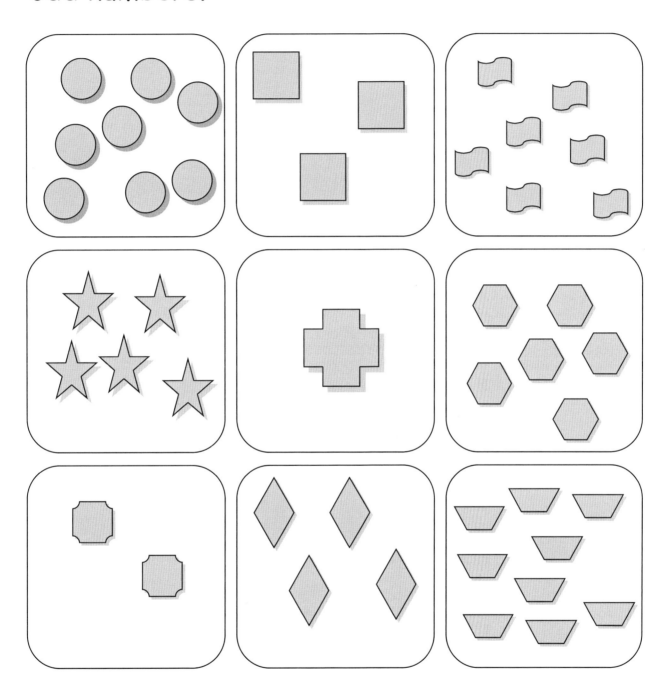

Concept: Understand even and odd numbers.
Introduction: Display two sets of ten or less objects, one containing an even number and one an odd number. Ask the child to put the objects in each set into pairs. Ask the child to count the objects in each set. Ask him/her which number is even. Tell the child that the set with an odd one out has an odd number of objects. Give the child a handful of objects and ask him/her to tell you if he/she has an even number or an odd number by making pairs.

Unit 31 — Even/Odd

Write the correct numbers in the boxes.

Odd numbers

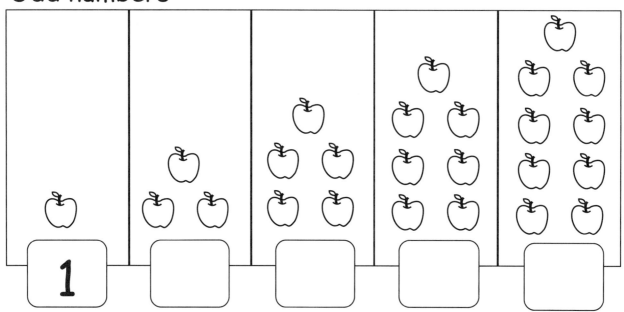

Even numbers

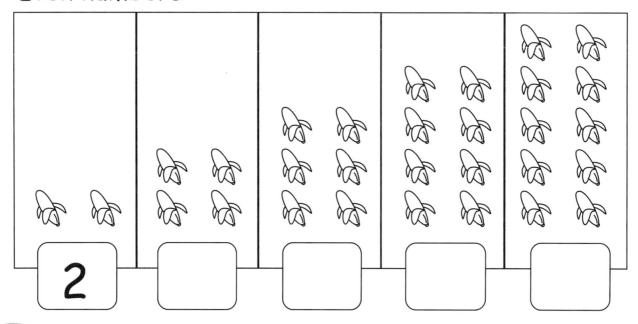

Unit 31 — Even/Odd

Count and write how many there are. Circle pairs. Circle the numbers that are even.

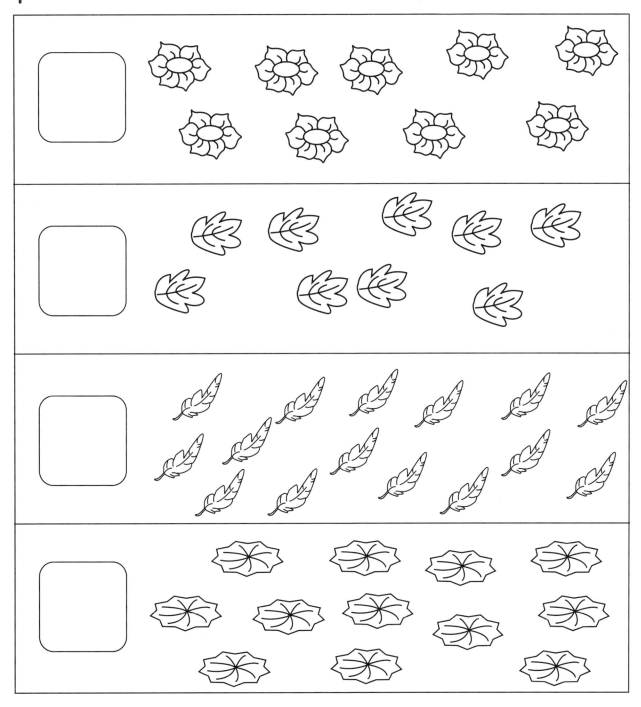

Concept: Understand even and odd numbers.
Introduction: Write a number between 1 and 20 and have the child use counters to determine if the number is even or odd. Repeat with other numbers.

Unit 31 — Even/Odd

Draw an even number of worms for the hen.

Draw an odd number of apples on the tree.

Draw an even number of buttons on the shirt.

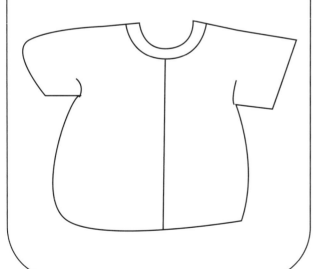

Draw an odd number of marbles in the bag.

Concept: Understand even and odd numbers.

Unit 32 — Fractions

Color the figures that have equal parts.

 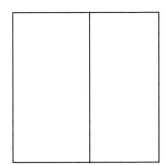 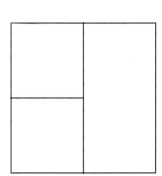

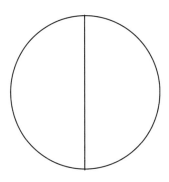 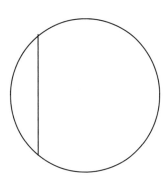 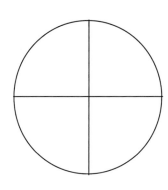

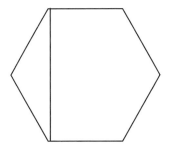 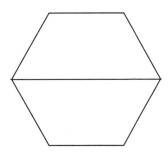 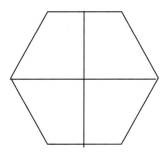

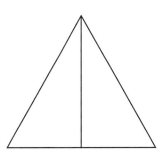 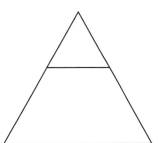 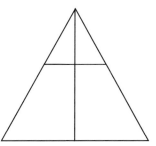

Concept: Identify equal parts.
Introduction: Give the child a piece of paper. Show him/her how to fold the paper to make two equal halves. Cut along the folded lines and then compare the halves to make sure they are equal. Repeat with fourths.

Unit 32 — Fractions

Divide the figures into 2 equal parts.

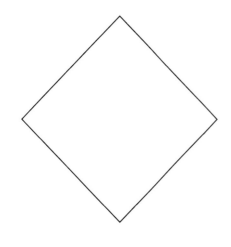
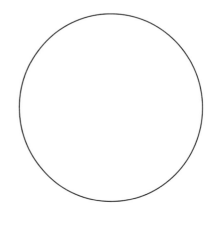

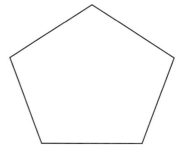

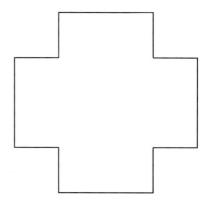

Concept: Make equal parts.
Introduction: Draw a shape and get the child to divide it into two equal parts. Tell the child that each part is a half.

Unit 32 — Fractions

Count the number of equal parts in each figure and write the number in the box. Color one part.

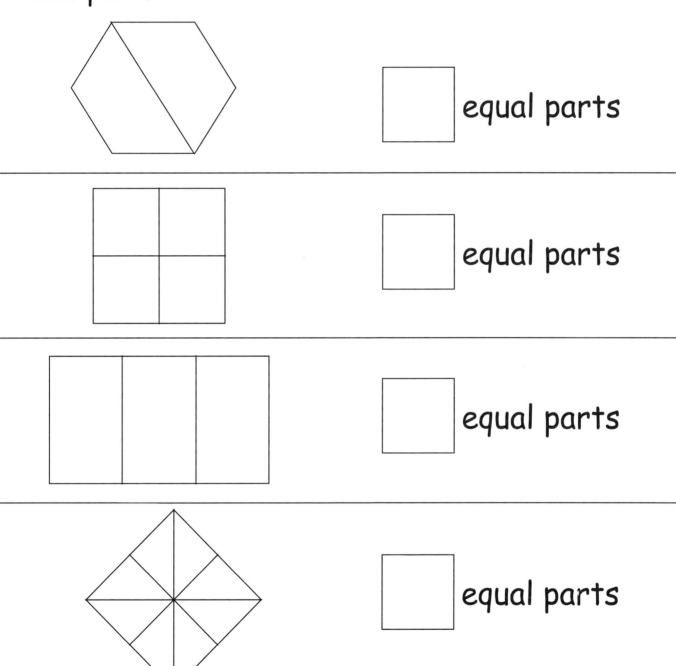

155

Concept: Understand a half and a fourth.
Introduction: Fold a paper strip into 4 equal parts. Get the child to count the number of parts. Ask him/her to color one part. Tell the child that one part out of 4 equal parts is called a fourth.

Unit 32 — Fractions

Color half of each figure.

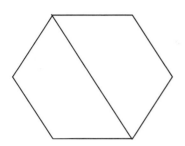
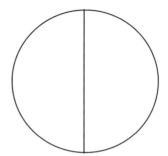
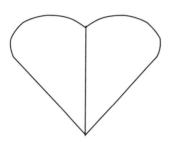
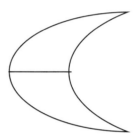

Color a fourth of each figure.

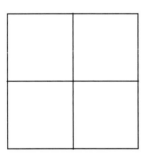
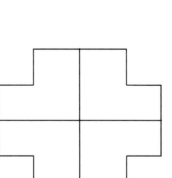
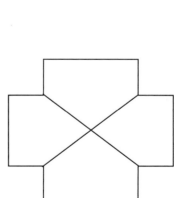

Concept: Understand half and fourth.

Unit 32 — Fractions

Color the correct number of parts.

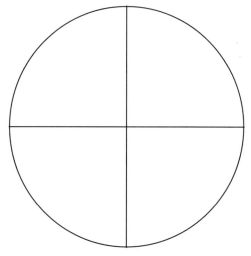

$\underline{1}$ part out of
4 equal parts

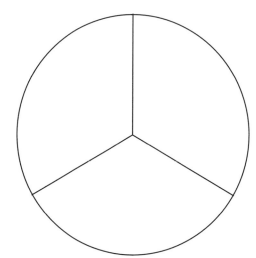

$\underline{2}$ parts out of
3 equal parts

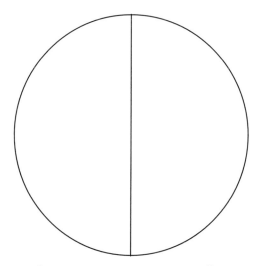

$\underline{1}$ part out of
2 equal parts

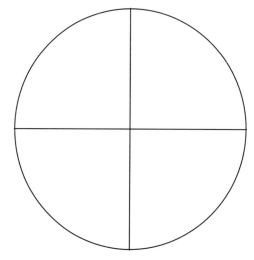

$\underline{3}$ parts out of
4 equal parts

Concept: Count equal parts of a whole.
Introduction: Fold a square piece of paper into fourths and ask the child to color 1 part out of 4. As you ask, write it as a fraction. Repeat with 2 parts out of 4, 3 parts out of 4, and 4 parts out of 4.

Unit 32 — Fractions

Color the correct number of parts.

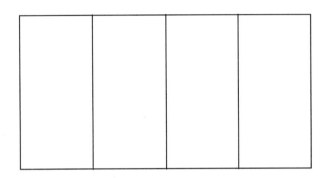

$\dfrac{3}{4}$ parts out of equal parts

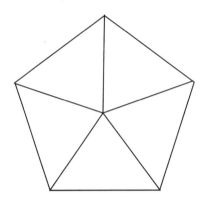

$\dfrac{2}{5}$ parts out of equal parts

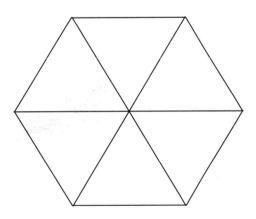

$\dfrac{3}{6}$ parts out of equal parts

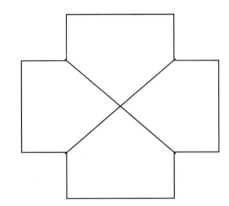

$\dfrac{3}{4}$ parts out of equal parts

Concept: Count equal parts of a whole.